GLACIER

The Earth series traces the historical significance and cultural history of natural phenomena. Written by experts who are passionate about their subject, titles in the series bring together science, art, literature, mythology, religion and popular culture, exploring and explaining the planet we inhabit in new and exciting ways.

Series editor: Daniel Allen

In the same series

Air Peter Adey
Cave Ralph Crane and Lisa Fletcher
Clouds Richard Hamblyn
Comets P. Andrew Karam
Desert Roslynn D. Haynes
Earthquake Andrew Robinson
Fire Stephen J. Pyne
Flood John Withington
Glacier Peter G. Knight
Gold Rebecca Zorach and Michael W. Phillips Jr
Ice Klaus Dodds
Islands Stephen A. Royle
Lightning Derek M. Elsom
Meteorite Maria Golia
Moon Edgar Williams
Mountain Veronica della Dora
North Pole Michael Bravo
Rainbows Daniel MacCannell
Silver Lindsay Shen
South Pole Elizabeth Leane
Storm John Withington
Swamp Anthony Wilson
Tsunami Richard Hamblyn
Volcano James Hamilton
Water Veronica Strang
Waterfall Brian J. Hudson

Glacier

Peter G. Knight

REAKTION BOOKS

This book is for C. G. Knight,
because I think he would have particularly liked this one

Published by Reaktion Books Ltd
Unit 32, Waterside
44–48 Wharf Road
London N1 7UX, UK
www.reaktionbooks.co.uk

First published 2019

Printed and bound in China

A catalogue record for this book is available from the British Library

ISBN 978 1 78914 134 4

CONTENTS

Preface

There are many ways of looking at the world, and what we see depends on how we choose to look. The artist, the scientist, the politician, the engineer: we each have different points of view, and so we each see different things. There are many books about glaciers that focus on glacier dynamics, climate change, geomorphology, physics, hydrology and so on. At first glance the physical sciences seem to have a virtual monopoly. This book does address some of that basic glacier science, including the history of the ice ages, ways in which glaciers create landscapes, and the role of glaciers in the big global system of climate change and sea-level rise, but this is not enough. Marcel Proust, in his novel *In Search of Lost Time*, wrote that 'the only true voyage of discovery consists not in exploring new places but in seeing through new eyes.' This book looks at glaciers through the eyes of explorers, politicians, artists, poets and storytellers as well as scientists. We learn by seeing things in different contexts, and we see more when we see from different perspectives.

Human life has developed in an unusual period of Earth's history: part of the 15 per cent or so of geological time during which glaciers have existed. But it is less than two hundred years since it was widely understood that glaciers come and go, that they were once more extensive than they are now, and that many of them are disappearing in response to human impact on the environment. This realization was a paradigm shift both in science and in our cultural relationship with nature. We can call it the glacial turn: we live not only in a physical ice age, in which

A mountain glacier descending into Glacier Bay, Alaska.

How we see glaciers depends on how we look at them, and our points of view have changed over time. This photo, taken from the International Space Station in September 2015, shows the glaciers of the Torres del Paine National Park in southern Chile, and a space cargo vehicle on its return journey to Earth.

glaciers affect the landscape, but in a cultural ice age, in which our understanding of how glaciers fit into the planet's life affects our whole view of the world and our place within it.

Increasingly, over the last two hundred years, glaciers have affected culture, spirituality and our view of ourselves. Those physical and cultural ice ages are reflected not only in our science, our environmental future and our economics, but in our art and our adventures. For readers who know little or nothing about glaciers this book provides a broad introduction ranging from the science of how glaciers work to the way that glaciers have featured in art and in the human imagination. For people who already know something about glaciers, I hope the book will offer some new points of view.

1 Ways of Thinking about Glaciers

A perennial mass of ice, and possibly firn and snow, originating on the land surface by the recrystallization of snow or other forms of solid precipitation and showing evidence of past or present flow.[1]

Today, from a distance, I saw you
walking away, and without a sound
the glittering face of a glacier
slid into the sea.[2]

Precisely defined for the scientist, a metaphor and an icon for the poet, glaciers for each of us have their own place in our view of the world.

Human life has developed in an unusual period of Earth's history: a period in which there are glaciers. Of the history of the planet, glaciers have been here, on and off, for only about 15 per cent of the time, but they have been here throughout the 2 million years of human history and prehistory. However, only for the last century have we lived in an age in which the importance of glaciers has been acknowledged. It is less than two hundred years since the idea was first widely accepted that ice ages come and go, and that glaciers used to cover much more of the Earth's surface than they do now. We have only just noticed that we are living in a glacial period. We live now, for the first time, in a cultural ice age as well as a physical one. Few of us consciously place 'glacier' at the heart of our view of the world but we do now appreciate the key role of glaciers both in the planet's past – creating the landscapes that surround us – and in its future, as major players in the unfolding drama of climate change. But at the same time that we noticed glaciers we also realized that the physical ice age may be coming to an end, that we are losing the glaciers, and that it may be our fault that we are losing them so fast.

Because we recognize the importance of glaciers in the global system we now have a particular view of the world, different from that of the generations before us. In glaciers we recognize nature's fragility, complexity, majesty, ephemerality, vastness, beauty, terror: in glaciers we see the sublime. Each of us is also touched by them very directly on a practical day to day level, wherever on the planet we live. Glaciers created my local English landscape, determining the nature of the soil and of the crops that grow here, but they also control the atmospheric and ocean circulations that drive weather systems across the globe. They influence sea level in every ocean, and they supply drinking and irrigation water to millions of people. Glaciers are not remote: they reach out to touch everything, everywhere. On the other hand, especially to lowland, mid-latitude city dwellers, the glacier is an alien creature whose name conjures images of wilderness, frozen wasteland and polar desert.

The way you think of glaciers depends on where you live, and when. At the Siachen Glacier, which reaches an altitude of 5,753 m (18,875 ft) on the disputed border between India and

Land, ice and ocean: glacier ice meeting the coast in Antarctica.

Glacier exotic: glacier-clad mountains of the Altai Range in Mongolia.

Overleaf: Close-up image of ice in a glacier, showing areas of clean ice (dark) and areas with air bubbles (light).

Pakistan, more than a thousand soldiers have died in recent decades in a long-running conflict waged at such a high elevation that far more casualties have been caused by avalanches and altitude-related illness than by gunfire. In 2012 an avalanche killed 129 Pakistani soldiers on the glacier. But the Siachen Glacier and its forgotten war do not feature prominently in the international headlines. For the vast majority of people, glaciers are far away or long ago. Most people have never seen a glacier. Glacier historian Mark Carey observed that some visitors looking up at the mountains in Glacier National Park in Montana, USA, cannot even recognize a glacier when they first see one.[3] By contrast, in some places glaciers play a major role in everyday life and in cultural history. There is a rich tradition of folk stories, myths and legends running through popular culture in the European Alps, and the Icelandic sagas are replete with glaciers and glacial meltwater floods. If you grew up with glaciers, live with glaciers,

and live in a culture rich in glacier mythology then you see them in a certain way. Anthropologist Julie Cruikshank, writing about the indigenous peoples of Alaska and the Yukon, describes their tradition of incorporating nature within human affairs:

> In . . . Athapaskan and Tlingit oral tradition, glaciers take action and respond to their surroundings. They are sensitive to smells and they listen. They make moral judgements and they punish infractions. Some elders who know them well describe them as both animate (endowed with life) and as animating (giving life to) the landscapes they inhabit.[4]

Today glaciers are most frequently encountered in popular culture under the heading of environmental protection. By contrast, for a period of several hundred years up until the mid-1800s glaciers were advancing in many parts of the world, causing devastating floods and avalanches, wiping out settlements and overwhelming populated areas. Brian Fagan called a chapter in his book about that period 'The War against Glaciers'.[5] It seems to be a war that we now regret having won so convincingly. Glaciers are disappearing from mountain tops around the world, causing a new set of environmental hazards and cultural changes.

From traditional to Romantic

Folk tales involving glaciers have been reported from many different locations. In the Ecuadorian Andes traditional stories explain the ice-capped peaks in terms of interrelationships between mythical characters represented by the mountains.[6] When the summit of Cotacachi is covered in snow, for example, one story tells us that it is a sign that her lover Imbabura (a neighbouring volcano) has visited in the night. When asked about climate change and the loss of ice from the 'darkening peaks' younger people with formal education talk about climate change, but older people refer back to events in traditional stories, and attribute the changes to Cotacachi's 'punishment'. Different

The Grinnell Glacier Trail, in Montana's Glacier National Park. The park's glaciers are retreating catastrophically, but the spectacular glacial landscape remains.

cultures, and different generations within the same culture, have different points of view. This contrast between tradition and enlightenment, between magic and science, can be traced back into the seventeenth century, as scientific explorers and Grand Tourists exported European attitudes and methods. The age of enlightenment, the age of romanticism, the age of exploration, the age of exploitation, the age of environmental conscience . . . wave after wave of re-visioning of landscapes.

In Europe before the age of enlightenment glaciers were viewed most commonly as remote, menacing and fearsome: an obstacle and a threat. Glaciers were associated with areas beyond

This picture from about 1903 shows a group of tourists on the Mer de Glace, Chamonix.

the reach of civilization or law, with bandits, avalanches and wild country. In this period glaciers were also an increasing menace: an advancing threat. During a period known as the 'Little Ice Age' between about the fourteenth century and the end of the nineteenth century glaciers across Europe and in other parts of the world experienced substantial advances. Land was lost beneath expanding glaciers, and damaging floods occurred when advancing ice temporarily blocked rivers, impounding lakes which subsequently burst with devastating effects.

By the end of the seventeenth century a new layer of subtlety was entering into European perceptions of glaciated and wilderness landscape. Increasing numbers of people were travelling through the Alps, many of them taking the 'Grand Tour' through Europe to the classical sites of Italy. En route they would pass through wild Alpine scenery, but would experience it from the perspective and with the physical and economic security of wealthy aristocracy. Today's equivalent, perhaps, is the 'Glacier Express' tourist railway, which runs between Zermatt and St Moritz and is described in its promotional literature as 'indescribably beautiful' for its 'roaring mountain streams and craggy cliff faces'. For the Grand Tourist, as today for the rail traveller, the 'fearful' landscape was, although close at hand, made safe to a level where the 'danger' could actually be enjoyed. Eighteenth-century philosophers such as Edmund Burke and Immanuel Kant developed the idea of the sublime to describe this feeling where the observer of wild and fearful nature might experience a quite pleasurable sensation from recognizing (at a safe distance) the overpowering strength of nature and our own insignificance. This idea flourished with the development of the Romantic movement in art and literature. Wilderness landscapes, mountains and glaciers, storms at sea and other natural terrors were presented by artists such as Turner and poets such as Wordsworth as evidence of nature's wild glory. As exploration of remote areas progressed, and civilized tourism penetrated further into formerly inaccessible areas, locations that could still be considered extreme wilderness and inspire sensations of the sublime became more and more remote. By the end of the nineteenth century the Alps were no

The Glacier Express en route to Zermatt.

This photograph from Roald Amundsen's 1910–11 Antarctic expedition was titled 'A striking view at the top of the Devil's Glacier, looking toward Hell's Gate'.

longer sufficiently extreme to engender adequate feelings of danger and awe in the popular imagination, and the polar regions started to take their place as the focus of our shared image of the ultimate fearful wilderness, our shared sublime.

The great polar explorations of the early twentieth century were founded in a fortunate coincidence of interest in exploration, imperialism and romantic adventure. The diaries and reports of Scott, Shackleton and others are self-consciously epic-heroic, clearly aware of the significance, quite beside all the science and discovery, of pitting humanity against the wilderness. The names of their ships set the tone: Scott sailed in the *Discovery*, Shackleton in the *Endurance*. Nansen sailed in the *Fram*, which is Norwegian for 'Forward'. The names of the features that they encountered and the role of glaciers as obstacles to progress have entered the language of popular culture. The Ross Ice Shelf is known as the Great Barrier. The Beardmore Glacier features prominently in Scott's diaries as a barely surmountable threshold to the polar plateau. Later in the century, the Khumbu Icefall on Everest, at the head of the Khumbu Glacier above base camp on the way to the summit, has taken on that same 'barrier' iconography – glacier as something to be crossed, surmounted, overcome on the way to a greater prize; glacier as fearful challenge; worthy adversary.

Science to the forefront

Scott's Antarctic expedition was about not only exploration and adventure, but science. Through the twentieth century right up to the present, as opportunities for genuine exploration of new

Climbers descending through the Khumbu Icefall down from Camp 1 to Base Camp on Mt Everest.

territory have diminished, adventurous expeditions into wilderness areas, whether genuinely motivated by science or in fact primarily adventure trips, have usually been at least partially justified by some scientific content. Even teams of schoolchildren taking trips to Iceland or Alaska seem to have to validate their requests for financial support with some claim to be 'doing science'. By the start of the twenty-first century, glaciers in popular imagination had become synonymous with concerns about global warming and were easy pickings for anybody to do some simple science in the context of environmental change. Glacier science has been closely involved in our developing understanding of climate change for nearly two hundred years. In the 1840s, Swiss scientist Louis Agassiz began to convince the geological community of what several people had already noticed: that glaciers used to be much more extensive in the past, and that this must have been associated with an 'ice age'. If there was an ice age in the past, and glaciers had subsequently retreated, that meant that the climate had changed. From that point onwards the sciences of climate change and glaciology progressed as close relations.

Scientists' tents on the surface of the Antarctic Ice Sheet during the 2002/3 U.S. National Snow and Ice Data Center 'Antarctic Megadunes' research programme.

Scientists and the public today are much better informed about glaciers than they were in the 1840s. Agassiz had no aerial photographs or satellite imagery. In 1840 we had no idea what lay in Antarctica or even in the heart of Greenland. Expedition reports were few and imperfect. Today, by contrast, even those of us far away from glaciers are kept well informed by torrents of information about threats to the environment, changing life-styles of remote mountain peoples, the break-up of another ice shelf or the latest change to the predicted date of disappearance of some glacier that had not even been discovered in 1840. Today on Google Earth we can find satellite imagery and photographs of glaciers on every part of the planet. The mystery of distance is being diminished. Heinz Zumbühl, a specialist from the University of Bern in the history of how glaciers have been measured and recorded, points out that 'the change of glacier representation techniques from drawings to photographs demonstrates also the changing view on glaciers from the magic to the scientific.'[7] Today's vivid and immediate streams of communication and shared experience did not exist for previous generations, and most people have lived their lives entirely untroubled by the notion of glaciers. What little we knew was shrouded in myth and imagination. Tall tales. But even today, despite the progress of science and the constant stream of news, there persists a great deal of ignorance of basic science, a great deal of unconcern on the part of huge numbers of people, and a disturbing amount of deliberate misinformation being produced as fallout from the environmental political battlefield. Unfortunate scientific blunders, such as that in 2007 by the Intergovernmental Panel on Climate Change citing 2035 as a likely date for the disappearance of Himalayan glaciers, generate publicity that masks real and important events such as the disappearance in 2009 of the Chacaltaya Glacier in Bolivia.

Although we know more about glaciers now than we did one hundred years ago, we still seem able to ignore the implications of what we know. For example, not many people outside Norway have heard of the glacier Jostedalsbreen, but it is a glacier so big it could supply Norway's water for one hundred years. Not many

A true-colour image of the terminus of the Bering Glacier, Alaska, acquired in September 2002 from the Landsat-7 satellite.

people outside Bolivia or the glaciological community noted the passing of Chacaltaya Glacier, but Chacaltaya was an early casualty in what might turn out to be a global environmental catastrophe, and its name should be a rallying call in the struggle to adapt to a changing world in which mountain glaciers are disappearing within human timespans.

We are only at the start of this book, but already we are talking about disappearing glaciers, environmental change and the social challenges that these might initiate. Today, conversations about glaciers turn quickly to talk of their disappearance. Writing about how we came to think of glaciers not just as an icon for global warming but as an 'endangered species', historian Mark Carey attributes our sense of loss to factors beyond science.[8] Conversations around glaciers and global warming are tied to conflicting ideologies about nature, imperialism, wilderness and global power. In other words, the idea that glaciers might be disappearing strikes a chord with a lot of different people for a lot of different reasons. Glaciers are tied to our environmental and social futures, and their fate is relevant to all of us. These are big, complicated issues. Science has not been the only response to them.

Art and imagination

Art and science are ways of making sense of the world around us: both make choices about what is important and both make choices about how to represent and communicate those matters of consequence. Science is not the only approach that people have taken to making sense of glaciers. Glaciers have inspired artists in different ways, from Albert Bierstadt's mountain landscape paintings and Mary Shelley's novel *Frankenstein* to the work of present-day artists such as Anna McKee, who has been motivated by collaboration with scientists to produce works based on the climate record contained in glaciers. For McKee, ice is a metaphor for the deep memory of the natural world and the frailty of its ecosystems. At the start of the twenty-first century the dominant refrain in the 'artist statements' that preface the gallery catalogues and websites of glacier artists is an

Chacaltaya glacier in the Bolivian Andes was the world's highest-altitude ski resort until the glacier disappeared in 2009.

environmental one. Artists are at the battlements. Artists such as McKee, Jill Pelto, Katie Paterson and members of the Cape Farewell expeditions have produced glacier-inspired art that both echoes our traditional imaginings and imbues them with a modern sensibility attuned to environmental awareness. There is an increasing tendency towards art-science collaborations, but this builds on a long-standing link between the science and the art of glaciers. Edward Wilson, who died with Scott on his last polar journey, combined science with art in his renderings of the Antarctic landscape. Modern climate science uses old paintings as data sources for measuring glacier retreat.

This book explores the differences, but also some clear similarities, between the artist's and the scientist's approach to glaciers. The comment 'they reveal an enquiring and analytical mind and a consistent exploration of the structure of landscape' was applied not to a scientific study but to the work of the twentieth-century British artist Wilhelmina Barns-Graham, who made a series of glacier drawings based on her travels in the European Alps.[9] English historian W. G. Hoskins wrote in 1955 that 'poets

Edward A. Wilson, *The Great Ice Barrier – Looking East from Cape Crozier*, 1911, watercolour, made during Scott's Antarctic expedition.

make the best topographers.'[10] In 2005 environmental author Siân Ede asked whether Science was the new Art.[11] The boundary between art and science is blurred, and it is becoming more so as glacier images are used as symbols for the environmental movement. For a TV producer, nothing but a polar bear standing on a small iceberg says 'global warming' more eloquently than film of ice crashing into the sea from the front of a glacier. On the environmental battleground the glacier has become poster child, and glacier art has become propaganda.

The glacial turn

This book's selection of examples of the different ways in which people have looked at glaciers is presented from a very specific perspective. Had I written the book a hundred years ago, or were it to be rewritten a hundred years into the future, its context and emphasis would be quite different. Scientists refer to major changes in the way we view the world as paradigm shifts. When Galileo put the Sun at the centre of the solar system, when Newton recognized gravity, or when the first astronauts saw Earth from space, our fundamental understanding of the world changed. In the arts and social sciences these major shifts in attitude are referred to as 'turns'. In literary studies Adriana Craciun and others have discussed an eighteenth-century 'Oceanic Turn'. In social sciences many research directions were replotted in the late twentieth century as part of a 'Cultural Turn'. Something similar can be recognized in the history of our understanding of glaciers. Prior to the mid-nineteenth century glaciers played a very small part in either the scientific or popular consciousness. Then from 1840 the widespread acceptance that glaciers had previously covered much of the planet and could one day expand to do so again led not only to a great upsurge in glaciological science but to a new popular awareness of the significance of glaciers. Charles Darwin published important work on glaciers. Glaciated territories featured in the heroic explorations of polar regions. Within a few decades glaciers and ice ages were at the heart of our understanding of how the planet works and of our place in the natural system. It was a paradigm shift, and we can call it the Glacial Turn.

The science aspect of this Glacial Turn is well documented, but the growing importance of glaciers in our cultural and psychological view of the world has been less well covered in the literature. Our understanding of glaciers developed remarkably suddenly, starting in the middle of the nineteenth century from effectively nothing when Louis Agassiz convinced the world that glaciers had once been much bigger and had changed the landscape in an enormous, ancient ice age. Our understanding of the physical world changed abruptly at that point, and so our perspective on our own context in the world also changed. The fact of glaciers and ice ages, our recognition of them, what I refer to as our 'noticing' of glaciers, in an almost metaphysical sense grounds our perception of our own place in the world. A world with glaciers in it gives us a particular recognition of both scale and fragility in the environment, and that recognition is reflected strongly in our image of ourselves within that environment. On the one hand glaciers and ice sheets make us feel very small. On the other hand, our impact upon them shows us to

Some glaciers are more accessible than others. This is Mount Sefton (Māori: *Maukatua*) in the Moorhouse Range of the southern Alps of New Zealand.

be very big in our ability to affect the planet. This evolving view has been reflected in the increasing sophistication of the ways in which glaciers have featured in art and, recently, in international environmental politics. We live, therefore, not only in a physical ice age (an age when there are glaciers present on Earth) but, and only for the last century or so, in a cultural ice age, in other words a period when humanity notices, recognizes and ascribes physical and cultural importance to glaciers. The tragedy of this cultural ice age is that the point at which we noticed glaciers was a point at which they were on the brink of collapse, and that their collapse may be due partly to our own actions.

Glaciers, the universe and everything

Tourists exploring the edge of Langjökull, Iceland's second largest glacier.

In their seminal 1976 book *Glaciers and Landscape* David Sugden and Brian John treated glaciers as 'whole systems', stressing the links between their 'functional components', with inputs, outputs, variables and balances.[12] It is an approach that is now central to the scientific study of glaciers and glacial landscapes, which are often treated as glacial 'landsystems'. We can expand that approach and consider the relationships, linkages and connections between glaciers and other aspects of our glacial experience. As a geographer by training I am always on the lookout for the 'Total Geography' in which the physical, social, artistic, scientific and emotional aspects of our environment are combined. Such cross-cultural mash-ups are increasingly common, and the world

The fractured surface of a glacier in New Zealand's Fiordland National Park.

of glaciers is rich in these nature-culture crossover experiences. For example, Glacier National Park in Montana has hosted both an artist-in-residence programme to stimulate artistic exploration of the landscape and a citizen science programme that uses trained visitors to collect scientific information. Special visitor weekends combine opportunities for history, science, art and outdoor activity.

People approaching glaciers from very different directions can easily arrive at common ground. Scientist Richard Alley famously described the understanding of how glaciers move as the holy grail for glacier science.[13] Mary Shelley in *Frankenstein* described 'the accumulated ice, which, through the silent working of immutable laws, was ever and anon rent and torn, as if it had been a plaything in their hands'. For both the scientist and the artist the laws of motion were at the heart of the glacier story. Artists and scientists frequently draw out the same key issues: time, movement and scale. These three are connected. The length of glacial time, the slowness of glacier motion and the enormity of glacier scale sit well together in the imagination. Accumulating slowly to a huge size, glaciers integrate time as mass, building up a ledger of century upon century of snowfall. Richard Alley uses the phrase 'two-mile time machine' to describe the cores of ice that can be drilled out of the world's ice sheets to reconstruct past climate.[14] Ice-core scientists are time travellers. Time is connected to speed: glacial age; glacial pace. And it is partly because ice moves slowly that it can build up into such huge masses that only gradually flow outwards and downwards to the oceans. The biggest solid object on the surface of the planet is the Antarctic Ice Sheet. In all that size, over all that time, ice sheets collect, accumulate, absorb and hide all those things that fall onto them from above and all those things that are drawn up into them from below – ancient volcanic dust, crashed aircraft, the bodies of lost explorers. The breathable atmosphere from 100,000 years ago caught up among crushed snowflakes is released on our tongues as we melt shards of ancient ice between our teeth. To the hydrologist a glacier is a storage point in the global water cycle. To a climate scientist the glacier is a store of ancient atmosphere. Geologists call glaciers conveyor belts, slowly carrying debris. To the glaciologist and geologist, the glacier is a purveyor and deliverer of ice, water and debris. The glacier also delivers history, a message encoded in the bubbles bursting out of the ancient ice – putting us in context, giving us inspiration, showing us who we are.

Glaciers have a particular place in the human imagination, and a particular status as a symbol or icon in fiction, poetry, music, painting and other representations. But the nature and complexity of that symbol is changing: glacier as creator of spectacular landscapes; water-supply resource; holiday destination; inspiration for creativity and spiritual reflection. For the environmental scientist the glacier is part of a global physical system and a barometer of global environmental change. It is a record book of the planet's climate history. At the same time glaciers can be seen as victims of human activity – an endangered species, an inspiration for environmental activism, a disappearing habitat. Then as the glaciers melt and pour themselves into the rising oceans, threatening coastal cities and depriving catchments of their irrigation water, glaciers become an environmental and political threat.

2 How Glaciers Work

A glacier is a body of ice that forms from the accumulation of snow on the ground and becomes thick enough to deform or flow under its own weight. People rework the official wording from time to time, but the idea of movement has always been central to definitions of glaciers. If ever a part of a glacier stops moving, glaciologists call it 'dead ice'. The etymology of the word 'glacier' refers specifically to a carrier or bringer of ice, but it is not only ice that glaciers carry with them: anything that falls onto or into or under a glacier can be entrained and carried along in what is one of nature's greatest conveyor belts – volcanic ash, pollen, meteorites. In 1942 a squadron of eight U.S. warplanes was abandoned after an emergency landing on the Greenland Ice Sheet. Fifty years later the planes were located 81.7 m (268 ft) below the surface, and one was excavated and restored. The remainder are likely to emerge at the edge of the ice sheet at some point in the distant future. In 1952 a U.S. Air Force C-124 Globemaster II crashed at Colony Glacier, Alaska, with the loss of all 52 passengers on board. The wreckage was quickly engulfed within the glacier. Moving forwards in the ice about a metre a day for sixty years, debris from the crash emerged 22.5 km (14 mi.) downstream at the front of the glacier where icebergs were taking wreckage with them into Lake George.

Over the duration of an ice age, rocks picked up by a glacier can be transported thousands of kilometres. These out-of-place rocks are called 'erratics' and they helped early geologists to work out the history of the ice ages. On a plinth outside my office in the sandstone country of central England is a granite boulder that was carried by glaciers from a specific outcrop of rock near Dalbeattie in Scotland, about 240 km (149 mi.) away. The boulder has a distinctive shape, with its sharp edges rounded off and with clear facets scraped into its sides, visibly reflecting the scraping and sliding of a rock that has been through this glacial conveyor belt along the bed of a glacier. In formerly glaciated areas, once you start looking for them these erratic boulders are everywhere. Because glaciers work precisely the way they do, they have a very specific and recognizable impact on the landscape.

The formation and survival of a glacier depends on the balance between the amount of ice being added by processes of accumulation (dominated by snowfall in most glaciers) and the amount lost by processes of ablation (such as melting or the calving of icebergs). That 'mass balance' determines a glacier's health. A positive mass balance (more accumulation than ablation) means a healthy, growing glacier. A neutral or zero balance means stability. A negative mass balance implies ice loss. Glaciers start to form when winter snowfall does not completely melt

Resting on the surface of Colony Glacier, Alaska, in 2015, is the landing gear from a C-124 Globemaster II aircraft that crashed in 1952. The debris was buried within the glacier for sixty years before beginning to emerge in 2012.

Boulders dragged underneath the ice for long distances and released at the margin of the ice sheet in West Greenland show the characteristic sub-rounded, faceted form of subglacially transported boulders. The bedrock shows the smooth surface characteristic of glacial abrasion.

away each summer but survives to be buried by the following winter's snow. As the snow builds up, the lower layers are compressed into ice. In environments such as Iceland where a lot of snow falls each year and the process is helped by some melting of snow and refreezing of the meltwater, the transformation of snow to ice can occur within a year or two. In colder, drier environments less snow falls, less water is present, and it can take hundreds of years. In the middle of Antarctica those layers of snowfall have built up to a depth of more than 4 km (2.5 mi.) of ice. We do not know for sure exactly how old the ice at the bottom is, but one ice core that drilled through 3,000 m (9,840 ft) of the ice sheet (European Project for Ice Coring in Antarctica (EPICA), Dome C) has extracted ice that fell as snow 800,000 years ago.

The Antarctic Ice Sheet is the largest of the planet's glaciers today, but glaciers come in all shapes and sizes. The biggest glaciers are the ice sheets, so huge that they completely bury the topography of the landscape beneath them and can cover whole continents. At present the Earth has only two ice sheets:

A large boulder carried on a glacier surface in Bylot Island. Debris such as this that might have fallen from the valley side and is carried on the glacier's surface retains its angular character and is largely unaltered by its journey on the ice, in contrast to debris carried at the glacier bed, which can be severely altered during transport.

one covering most of Antarctica and one covering most of Greenland. The Antarctic Ice Sheet is so thick that entire mountain ranges are buried within it and leave no trace at the surface. The Gamburtsev Mountains in East Antarctica cover an area similar to that of the European Alps, and reach almost to the same elevation, but are completely submerged beneath Antarctica's thick ice. A rift valley as deep as the Grand Canyon that nobody had noticed before was discovered beneath the Antarctic Ice Sheet in 2012. A valley a mile deep, completely lost beneath the ice, was revealed eventually by scientists using ice-penetrating radar to try and find out why some parts of the ice sheet were melting more than others. Then in 2016 an even longer canyon system was reported, and geologist Stewart Jamieson declared that the bed of Antarctica is less well known than the surface of Mars![1]

In the past there have been other ice sheets. Until about 12,000 years ago the Laurentide Ice Sheet, covering most of what is now Canada and the northern portion of the USA, was substantially larger than the Antarctic Ice Sheet is today. When the Laurentide Ice Sheet melted it released a volume of water equivalent to a depth of more than 70 m (230 ft) across the world's oceans: a global flood of biblical proportions. In fact, the flood at the end of the last ice age probably lies at the heart of the flood myths of many cultures, which may be more legend than myth. Between them, the Greenland and Antarctic ice sheets today contain about 99 per cent of the volume of all the world's glacier ice, with 91 per cent of it in Antarctica, 8 per cent in Greenland, and the remainder spread among all the smaller glaciers. Some scientists treat ice sheets as completely different creatures from all the other types of glacier, even to the extent that there are books about 'Glaciers and Ice Sheets' as if they are two different things. However, even though they have special impacts on the planet's environment, ice sheets are really just huge glaciers.

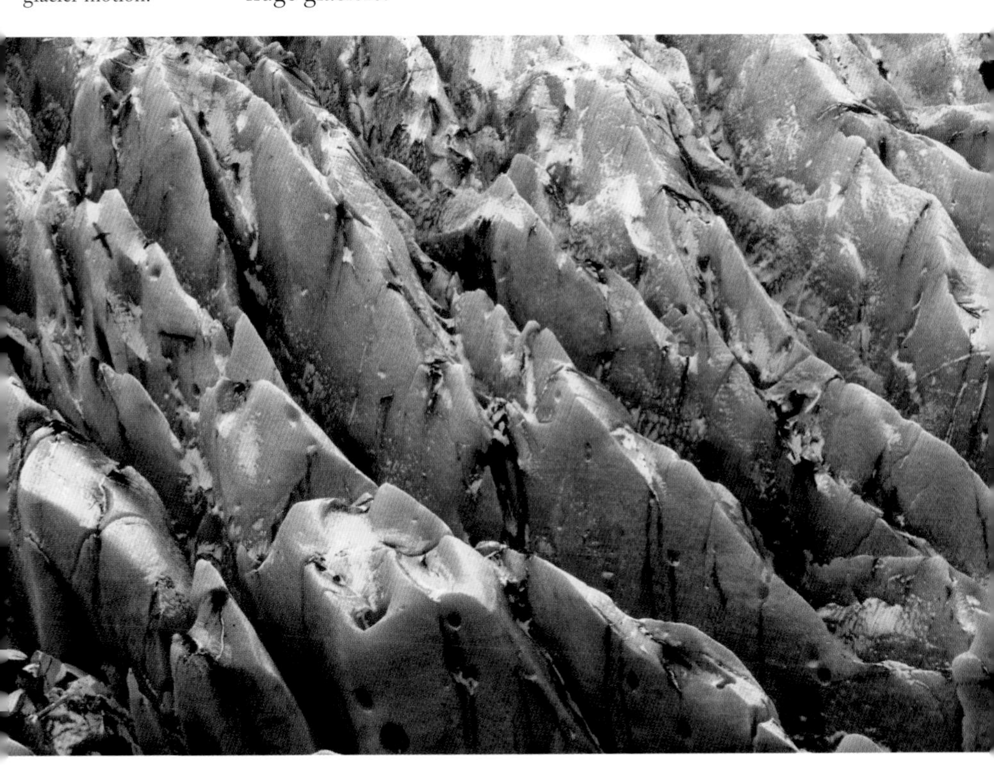

Crevasses on the glacier surface reflect fracture of the ice in response to stresses associated with glacier motion.

There is a whole bestiary of glacier types. Much like ice sheets, but on a smaller scale (technically defined as being less than 50,000 square kilometres but still overwhelming the topography beneath them) are ice caps such as Mýrdalsjökull in Iceland. Also covering significant areas but not completely burying the topography beneath them are icefields, which drape across mountain ranges letting the peaks show through and allowing their ice to be steered through the valleys. One of the best known of these is the Columbia Icefield in the Canadian Rockies, conveniently accessible by a road that is advertised as 'The Icefields Parkway: The world's most spectacular journey'. Sometimes an individual mountain has its own local drape of ice known as a carapace, such as those that provide the icy topping to volcanic peaks like Chimborazo and Cotopaxi in the South American Andes, or the famous and ill-fated glaciers that have previously covered mountains such as Kilimanjaro, but which may not do so for very much longer as the Earth's environment continues to change. The long glaciers that stretch outwards from the great

The Elephant Foot Glacier, in Greenland, is a piedmont glacier lobe, emerging through a gap in the mountains and expanding onto the flatter valley floor.

Since the mid-1970s members of the ANSMET (Antarctic Search for Meteorites) team have recovered more than 20,000 meteorites from the ice surface in Antarctica. Here a new find is examined in December 2015.

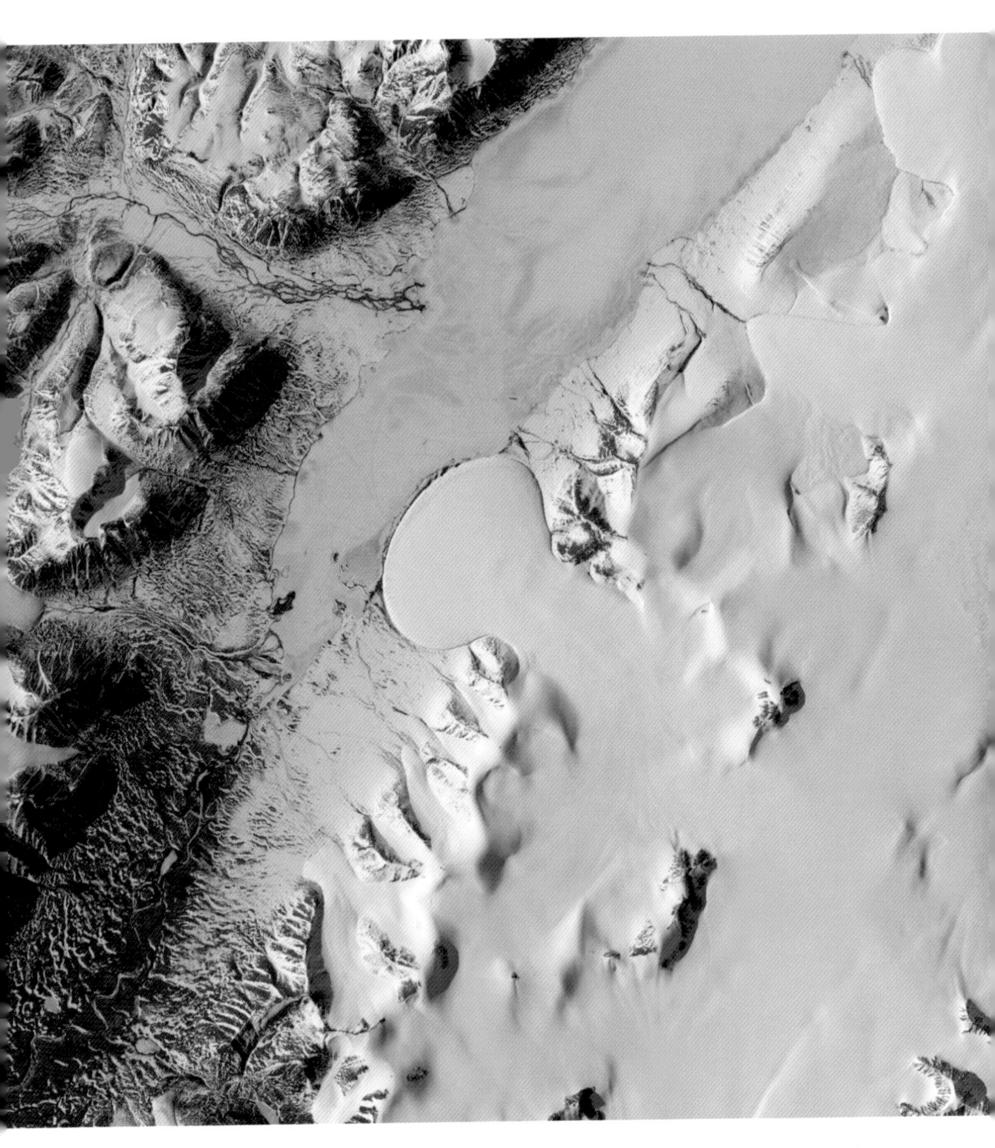

ice sheets, or grow from their own local mountain accumulation areas and snake their way through the topography like rivers of ice are called valley glaciers. The Fedchenko Glacier in Tajikistan, commonly listed as the longest glacier outside the polar regions, is around 80 km (50 mi.) long. Isolated high on mountainsides, tiny glaciers known variously as 'cirque' or 'corrie' glaciers occupy

hollows eroded into the slopes. The smallest are known as 'niche' glaciers, clinging to mountainsides as patches of flowing ice less than a square km in area. They are so numerous, and in such a variety of remote locations, that even in this age of satellite technology there is no reliable map to show where all the world's small glaciers are. The so-called 'First Complete Glacier Inventory for the Whole of Greenland', published in 2012, was based on satellite data collected between 1999 and 2002, so it was already ten years out of date when it first appeared.[2] Even if there were a complete and current map of all the world's glaciers, the rate of environmental change is such that it would be wrong next year. With glaciers more than with most features of the natural environment, it is always possible to make new maps, to discover new things about the rapidly changing world, and then to find that they have changed again!

A variety of different glacial environments are visible in this image of part of the Penny Ice Cap on Baffin Island, Nunavut, Canada, including the ice cap itself and an outlet glacier with several tributaries.

Ice moves by three main mechanisms: sliding across its bed; internal deformation (sometimes known as creep); and deformation of the material on which the glacier rests. Sliding seems easy to imagine if you play with an ice cube on a table top, but the real world is not as smooth as a table top and glaciers have to negotiate the roughness – whether it be at the scale of millimetre bumps or mountains – in a variety of ways. In glaciers where the ice is close to the melting temperature, movement around obstacles can be accomplished by melting on the upstream side and refreezing on the downstream side, but in cold glaciers this doesn't work so well. That is one of the main ways that glaciologists differentiate between glacier types: warm or temperate glaciers are close to the melting temperature; cold glaciers are not. Many glaciers are cold in some parts and warm in others, and we call them 'polythermal'. On the whole, cold glaciers do not move so fast as warm ones. They don't slide so well if they are frozen to their bed, and they don't creep so well

The volcano Chimborazo is close to the equator, and owing to the Earth's equatorial bulge its summit ice cap is the farthest glacier from the centre of the Earth.

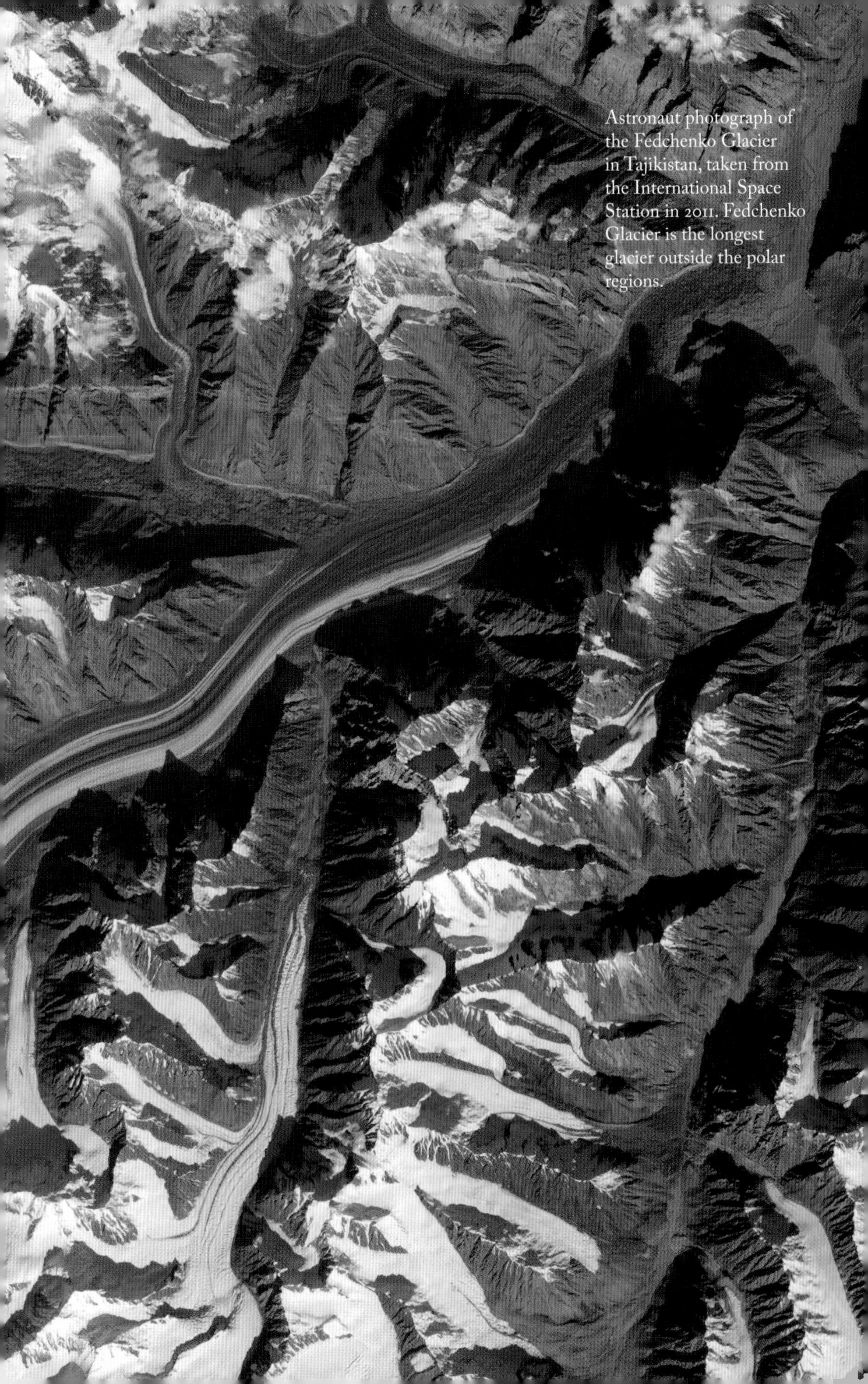

Astronaut photograph of the Fedchenko Glacier in Tajikistan, taken from the International Space Station in 2011. Fedchenko Glacier is the longest glacier outside the polar regions.

Water flowing at the base of the glacier Sólheimajökull in southern Iceland emerges through a spectacular portal in the ice front.

Water in glaciers can follow complex pathways over the surface, through the interior and at the base of the ice. Fountain Glacier, Bylot Island.

either. Creep, or internal deformation, is the process by which the ice squeezes along under the pressure of its own weight by mechanisms such as rearranging its crystal structures. Colder ice is stiffer than warmer ice and does not deform so easily. Some glaciers rest on rigid bedrock, but many rest on beds made of loose sediment and these can move by applying force to that underlying sediment so that it moves along underneath them, carrying the glacier with it. This idea of subglacial deformation was not taken into account much before the late 1970s, but for many glaciers it is the major part of how they move. All these types of movement vary depending on whether the ice and bed are close to the melting point, and the melting point depends not only on temperature but on pressure. At regular atmospheric pressure on my table top my ice cube melts, and water will freeze, at 0°C (32°F). However, at the higher pressures that occur underneath glaciers that critical temperature can be a little bit lower, so water can stay liquid, and ice can melt, just a little below 0°C. Because pressure underneath glaciers changes both geographically and over time, for example as glaciers get thinner or thicker, their beds can go from being 'cold' to being 'warm' and back again even without a change in temperature. Changes in ice thickness also change the ability of the ice to conduct geothermal heat away from the glacier bed, so thicker glaciers tend to keep the ground beneath them just a little warmer than thin glaciers. This can have a massive effect on glacier movement, and movement, in turn, can have a massive effect on the size and shape of a glacier.

The Laurentide Ice Sheet, which covered much of North America during the last ice age, went through a series of what

At the base of some glaciers, debris from the bed can be entrained into the glacier to create a debris-rich basal ice layer such as at this location at the Russell Glacier in West Greenland.

Many glaciers, such as Fountain Glacier on Bylot Island in this image, carry substantial amounts of debris within the ice.

A hiker in the Wrangel-St Elias National Park in Alaska walks along the crest of a glacial moraine ridge. The glacier that left this lateral moraine on the valley side can be seen on the right of the image.

have been called 'binge-purge oscillations', which illustrate how serious these effects can be.[3] For periods of thousands of years at a time the Laurentide Ice Sheet moved mainly by sliding over its bed, and moved at a fairly typical 'glacial' pace. However, from time to time, about every 7,000 years, the ice seems to have gradually become so thick that parts of the bed started to melt and the frozen sediment underneath the glacier became mobile. Subglacial deformation typically allows glaciers to go faster than sliding, so the ice sheet sped up. Because it was moving faster, a bit like warmed-up toffee, the ice spread itself thinner and spread out further, so the edges of the ice sheet extended rapidly. Ridges of debris known as moraines, pushed up and dumped at the edges of the advancing ice, mark positions to which the ice sheet expanded during these surge-like advances. In the northeast, the flow took the ice sheet through Hudson Bay and the Hudson Strait and caused the release of armadas of icebergs into the North Atlantic ocean. As the icebergs floated across the ocean they gradually rained out debris that the glacier had picked up across Canada. Sediment cores drilled out from the ocean bed over the last few decades include layer after layer of this iceberg-rafted debris. They are often called 'Heinrich layers' and they feature prominently in scientific work studying the history of climate change. Each time the ice sheet spread out and icebergs were released, the ice sheet became thin, the pressure and temperature at the bed was reduced and the bed lost its mobility. Flow returned to normal sliding, the ice slowed down, the flux of icebergs into the Atlantic stopped and the ice slowly started to get thicker once more in the centre. Time passed. Eventually, the ice in the centre reached such a thickness that the basal thaw began again, and another cycle of bed deformation, glacier advance and iceberg production started. The release of all those icebergs and meltwater into the ocean had a serious effect on the climate. The jury is out on whether the Antarctic Ice Sheet might be susceptible to similar extreme fluctuations, whether those might be initiated by our current phase of global warming, and what kind of climate impact they could have.

A similar cyclic variability, known as surging, occurs at smaller scales in many glaciers. Most glaciers typically move at a rate of somewhere around a few millimetres or centimetres a day, but some glaciers go through occasional periods of much faster movement where they can move at speeds of up to several metres per hour. The mechanism for surging can be based on subglacial deformation, like the binge-purge oscillations of the Laurentide Ice Sheet but on a smaller scale, or on the opening and closing of meltwater tunnels under the ice. One of the best-known examples of a surging glacier is Variegated Glacier in Alaska. After surges had been noticed at regular intervals through the twentieth century, another surge was predicted to occur in the early 1980s, and a huge scientific effort went into measuring and monitoring the 1982–3 surge to try and find out what might cause these events. It was found that when the ice reaches a critical thickness the meltwater tunnels at the base become closed off and the pressure of the trapped water floats sections of the glacier off the bed, reducing the friction and creating a period

Lowell Glacier in Kluane National Park, Yukon. The dark stripes on the glacier surface are 'medial moraines', lines of debris being transported by the glacier. The glacier's reduced thickness is indicated by the prominent 'trim line', or vegetation limit, on the valley side.

of rapid flow. The rapid flow causes the glacier to thin, normal drainage resumes, and there is a period of normal flow while the glacier slowly returns to its pre-surge thickness and the process repeats.

As the size and shape of an ice sheet changes over time, its direction of motion and its effect on landforms also changes. Directions of flow of former glaciers can be reconstructed by looking at scratches, grooves and ridges that they carve into bedrock or mould in soft sediment. Studies of ancient rock surfaces over tens of thousands of square kilometres of Canada have revealed complex overlapping patterns of these linear features, indicating different flow patterns in the ice at different dates.

Tiny grooves known as striations, scratched onto the bedrock by fragments of debris being dragged along at the base of the ice, are among the smallest examples of a wide range of different landforms that can be used to infer the presence and behaviour of ancient glaciers. We have learnt to interpret many of these landforms mainly by studying them at present-day glaciers where they are being actively created. For example ice-marginal moraines in Greenland or Iceland give us a perfect

Striations, grooves and other marks on bedrock indicate the passage of ice across this surface in Mount Rainier National Park.

Valleys carved by glacial erosion often have the characteristic U-shaped cross-profile demonstrated here by Lake McDonald Valley in Montana's Glacier National Park.

living case study to help us understand ancient moraines left behind by ancient ice sheets in areas from which the glaciers have long since disappeared. However, some landforms are actively created only underneath the ice where we cannot see what is going on. One of the most enigmatic of those is the drumlin, a kind of small rounded hill that has become the classic example of a controversial landform. At extreme ends of the spectrum of debate are those who argue that drumlins are created by huge subglacial 'megafloods' of meltwater that erode the ground beneath the ice, leaving these erosional remnants behind, and those who argue that the drumlins are deposited when a glacier is melting at its base and sets down its load of ice-transported debris. In the increasingly popular middle ground is a developing consensus that while different situations probably apply to different cases, the most widespread origin of drumlins is in the deformation of sediment beneath a glacier, such as in those situations mentioned earlier, when the bed becomes mobile and the ice can move more easily over a soft, deforming layer of subglacial debris.

In this aerial image of a drumlin at Raderach, in southern Germany, the pattern of houses built in the raised land of the drumlin clearly shows the characteristic streamlined form of these glacial hills, with the blunt end pointing upstream against the direction of ice movement.

Drumlins often occur together in huge numbers, or 'swarms', revealed by satellite imagery that shows the impact of large-scale flow patterns within ice sheets. Many linear features such as elongated swarms of drumlins and extensive suites of grooves and ridges are associated with the tracks of ice streams. Ice streams are zones of fast-flowing ice within the slow-flowing mass of an ice sheet: rather like swifter currents flowing through an ocean. Major ice streams in the present-day Antarctic and Greenland ice sheets include the ice streams that flow into the Ross Ice Shelf in Antarctica and the Jakobshavn Glacier ice stream in West Greenland. The existence of an ice stream in the southwestern section of the last British-Irish ice sheet probably explains how a lobe of ice was able to extend as far south as the Isles of Scilly while other sections of the southern margin of the ice sheet terminated much further north.

Drumlins are fascinating in their way, and impressive in their swarms, but not individually spectacular. But the larger glacial landforms, such as U-shaped valleys, pyramid peaks, fjords and landscapes made up of assemblages of these landforms, are among the most spectacular in the world. The Yosemite Valley in California, the Matterhorn in Switzerland and Milford Sound in New Zealand are among the Great Landscapes that we can attribute to the activity of glaciers. The Jungfrau-Aletsch UNESCO World Heritage site in the Swiss Alps was designated partly on the grounds of its glaciers and glacial landscapes, and especially the way in which its landscape has played an important role in European art and literature, as well as mountaineering and Alpine tourism.

For most people it is the spectacle of landscape, not the fine print of science, that draws us to glaciers. Even for many scientists, their way into the discipline was through a love of the great outdoors, or adventure, or spectacular landscapes. But the science and the scenery are connected. Science is only one way of looking at the scenery, but as a starting point for artists, economists or tourists, a basic appreciation of the science behind landscape helps us not only to understand but to better appreciate some of

The area around Switzerland's Matterhorn, a classic example of a glacial landscape.

The Great
Aletsch Glacier.

its aesthetic qualities. The physicist Richard Feynman famously said that far from undermining our aesthetic appreciation of nature, scientific knowledge only adds to the excitement, the mystery and the awe:

> There are often great distances between the detailed laws and the main aspects of real phenomena. For example, if you watch a glacier from a distance, and see the big rocks falling into the sea, and the way the ice moves, and so forth, it is not really essential to remember that it is made out of little hexagonal ice crystals. Yet if understood well enough the motion of the glacier is in fact a consequence of the character of the hexagonal ice crystals. But it takes quite a while to understand all the behaviour of the glacier (in fact nobody knows enough about ice yet, no matter how much they've studied the crystal). However the hope is that if we do understand the ice crystal we shall ultimately understand the glacier.[4]

3 Ice Ages: Glaciers Come, Glaciers Go

Earth's history has been marked by repeated changes of climate including, at the grand scale, shifts between periods of many millions of years during which the planet has at least some glaciers, and periods of many millions of years during which it does not. Within each of those extended glacial and interglacial periods the climate changed on shorter time scales, as glaciers came and went, expanded and contracted, in what we call stadials (short periods with more glaciers) and interstadials (periods with fewer or no glaciers). The more closely we look at the history of glaciers the more comings and goings we recognize, but the further back into history we look, the less clearly we can see. Our picture of the most recent 'ice age', which has lasted for a few million years and is still going on, is quite detailed. For the last few hundred years we even have paintings, written records and photographs as well as evidence in the rocks, the landscape, and the ice itself. By contrast, our picture of the glaciation that occurred in the Archaean nearly 3 billion years ago is quite sketchy, based on mere scraps of geological evidence. Ice sheets erode and destroy much of the geologic evidence of previous glaciations each time they readvance, making it hard to reconstruct their history beyond the last event and forcing geologists to be ever more inventive in their interrogations of subtle clues. Our knowledge has grown as new techniques for environmental reconstruction have been developed. In the 1840s geologists argued passionately about whether there had ever been even one ice age. By the early 1900s it was widely accepted that there had been four. In the 1970s, with

An ancient glacial till from the Neoproterozoic era, about 750 million years ago, preserved as a metatillite (a lithified till that has been affected by pressure or temperature), in Virginia, USA.

new evidence from deep ocean sediments and polar ice sheets, we recognized that there had been perhaps sixteen distinct fluctuations in the last 1.6 million years alone. The most recent literature is referring to as many as 45 discrete cold episodes during the present 'ice-house' period of the last 2.6 million years, and four or five previous ice-house periods in Earth's history. The Earth is approximately 4.5 billion years old, and its history of alternating glacial and interglacial periods stretches right back through that immense span of time.

Early Ice Ages, billions of years ago . . .

Very little geological record survives from the first billion years or so of Earth's history, a period called the 'Hadean' that was characterized by a molten surface slowly cooling to make the Earth's first rocky crust. The oldest signs of glaciation are rocks in South Africa that were laid down as glacial deposits in the 'Archaean' era about 2.9 billion years ago. At this stage in the history of the solar system the Sun's output was significantly lower than at present, and the simplest explanations for variations between glacial and non-glacial conditions under this faint young sun invoke changes in the amount of greenhouse gases such as carbon dioxide and methane in the atmosphere. When concentrations of these gases were relatively high the atmosphere would keep the surface of the planet warm, but when concentrations were low the temperature would drop sufficiently to allow glaciation. Variations in these gases occurred as the Earth's young surface evolved, oceans emerged and the continents started to form. For example, the development of a rocky surface crust created the first opportunity for chemical weathering of rocks to occur, and weathering involves processes that can 'draw down' carbon dioxide from the atmosphere and trap it in rock. Carbon dioxide combines with rain to make a weak carbonic acid, which falls to the ground, attacks minerals and creates, as by-product, carbonate rocks which essentially lock up the original carbon dioxide from the atmosphere. The removal of carbon dioxide from the atmosphere would have had a knock-on effect on atmospheric methane, making it more likely to condense into clouds, which would both reduce its effectiveness as a greenhouse gas and also cut down the amount of solar radiation able to reach the Earth's surface. It has been suggested that the likely timing of these processes early in the Earth's history could have resulted in depletion of atmospheric greenhouse gases sufficient to initiate glaciation about 2.9 billion years ago, the point at which we see the first slight evidence of glaciation in the geological record.

We have slightly more evidence for what appears to be the next glacial period, around 2.3 billion years ago, in what is known

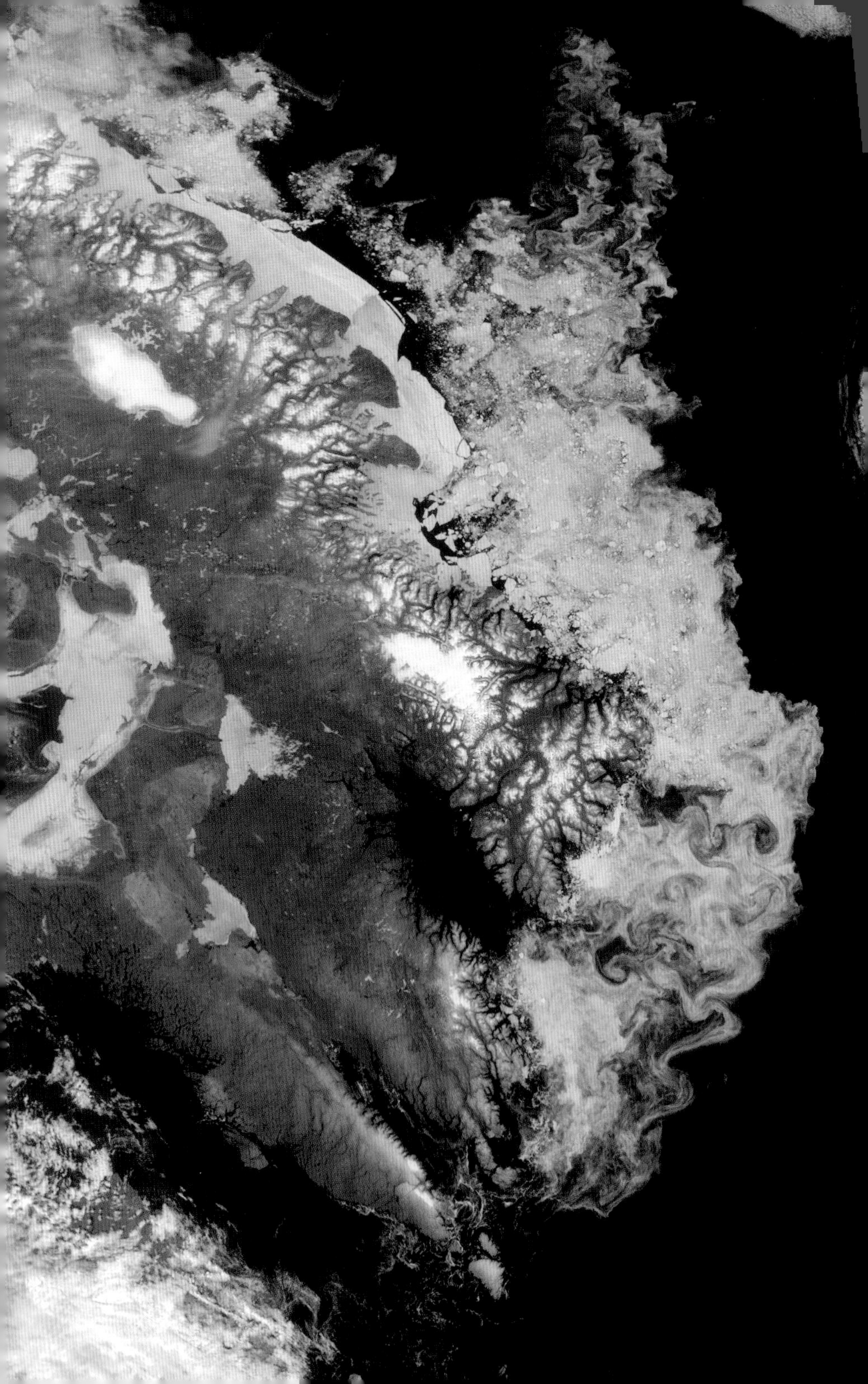

A thin section or slice of ice from an ice core drilled into the surface of the Antarctic Ice Sheet, in the Patriot Hills, Antarctica.

as the 'Palaeoproterozoic' era. For this we have glacial deposits in North America, Africa and Australia. As Earth's early atmosphere evolved, its oxygen content increased markedly about 2.5 billion years ago. This was due in large part to the production of oxygen by photosynthesis by early life forms. Initially, the oxygen they produced was taken up by iron in the environment and so did not enter the atmosphere (instead contributing to banded iron formations that are prominent in the geological record). From about 2.5 billion years ago, however, the available iron was largely depleted and oxygen produced by photosynthesis was released to the atmosphere rather than being locked into rocks. This oxygenation of the atmosphere from around 2.3 billion years ago had the effect of removing much of the atmosphere's methane, and hence, once again, reducing the atmosphere's greenhouse effect and initiating glaciation.

Barnes Ice Cap on Baffin Island is thought to be a surviving remnant of ice from the ice-age Laurentide Ice Sheet.

And then, 750 million years ago . . .

After those Palaeoproterozoic glaciations there followed a period of more than a billion years, sometimes called 'the boring billion', for which evidence suggests a warmer climate and no glaciations anywhere on the planet until around 750 million years ago (in the Neoproterozoic era), when the Earth cooled again and we have clear evidence once more of widespread glaciation. Glaciers extended down to sea level even in tropical areas and it has been argued that ice cover was so extensive that the whole planet became frozen. This is known as the 'Snowball Earth' hypothesis: every continent was covered by snow and ice; every ocean was frozen. The global mean temperature was about -50°C (-58°F), and it was about -20°C (-4°F) even at the equator. This Snowball Earth situation could arise as a result of runaway albedo feedback: as glaciers extended, the Earth's surface became more

Glaciation during the Permian period, 290–250 million years ago, deposited this till, containing granite erratics, in the Inman Valley south of Adelaide in south Australia.

reflective, causing the Earth to cool further, leading to greater glaciation, further increases in reflectivity, further cooling and extending glaciation until the entire planet was icy. Another explanation that has been put forward is that the break-up of a supercontinent called Rodinia resulted in the creation of lots of new continental margin areas which were sinks for organic carbon and would have led to depletion of atmospheric CO_2, reducing the greenhouse effect and cooling the Earth. The Snowball Earth frozen lockdown could have eventually been ended by volcanic activity that changed the greenhouse composition of the atmosphere and changed the reflectivity of the planet's surface by dusting the snow and ice with dark ash. Not everybody accepts that a Snowball Earth is even possible, let alone that it actually happened. One alternative model to explain the geological evidence is the so-called 'slushball Earth', where there remained unfrozen sea surfaces. Another is the idea of low-latitude glaciations with ice-free polar areas caused by extreme tilting of the Earth's axis, which would have reversed the balance of solar heating between the poles and the tropics. This controversial and exciting area of glacial geology highlights the interconnection of ice, rock, ocean and atmosphere.

After the Neoproterozoic there followed another long spell of warmer, ice-free conditions stretching for more than 100 million years, and then as we move into the more familiar part of the geological column we see a series of glaciations spread across the late Ordovician and Silurian periods about 450–420 million years ago and then another in the Devonian, Carboniferous and Permian around 375–263 million years ago. Interglacial-glacial transitions seem to be controlled by shifts from a high-CO_2 atmosphere to a low-CO_2 atmosphere, driven either by the evolution of plant species (particularly for the Permo-Carboniferous cold period at about 300 million years) or by changes in rock weathering driven by geological events. Our reconstructions of those ice ages several hundred million years ago are clearer than those of the ice ages billions of years in the past, but they are still hazy compared with our appreciation of the history of our current ice age, the so-called 'Late Cenozoic' glacial era.

South of Adelaide in south Australia, Selwyn Rock, named after the government geologist who first recorded it, shows grooves and striations carved by ice during the Permian era about 270 million years ago.

Just recently, 50 million years ago . . .

At the start of the Cenozoic era, 65.5 million years ago, the world was ice-free, but from about 50 million years ago there was a prolonged and significant cooling. One theory for the onset of this cooling involves, yet again, drawdown of CO_2 from the atmosphere. This time the drawdown was caused by a prolific blooming in the Arctic of the freshwater fern *Azolla*, which is particularly effective at taking on atmospheric CO_2. *Azolla* is preserved in significant amounts in a fossilized layer across the Arctic basin, indicating a period of perhaps 800,000 years when this plant was abundant and locked up CO_2 in ocean sediments rather than returning it to the atmosphere. That substantial disturbance to the carbon cycle may have been enough to initiate this latest ice-house era. The first trickles of geological evidence for the start of the transition from non-glacial to glacial conditions appear between 50 and 30 million years ago. The progressive cooling was translated into glaciation in different parts of the world at different rates. About 25 million years ago Australia moved apart from

Antarctica, and a circumpolar ocean current effectively isolated the southern continent from the warming effects of more equatorial currents. At the same time, Antarctica was positioned at the pole, meaning that snow and ice could build up on a land mass, developing into the kind of thick ice sheet that would not be likely if the pole were oceanic. By about 14 million years ago the Antarctic Ice Sheet was fully established. In the northern hemisphere, with no land mass at the pole, conditions were less favourable for glaciation, but around 3 million years ago the closure of the gap between North and South America at the Isthmus of Panama led to a change in northern hemisphere ocean currents, bringing more moisture to the Arctic and hence more snowfall to northern latitudes. By 2.4 million years ago substantial ice sheets existed in the northern hemisphere as well as the southern.

Mt St Helens, USA, showing the new glacier formed inside the crater since the eruption of 1980.

Right now: the last 3 million years

There is evidence of a big change in the nature of the climate system and glacial history at some point between 3 million and 1.5 million years ago. The exact date remains controversial but it is now usually placed at 2.6 million years ago. The change might have been caused by climate feedbacks initiated by the formation of ice sheets at both poles, or it might have been due to the particular configuration of oceans, continents and mountain ranges and the way they affected atmospheric circulation, but from that point onwards the climate system was, and continues to be, characterized by more frequent and substantial fluctuations. Glaciers have come and gone with more frequency and intensity in the last 2.6 million years than they ever seem to have done previously. This switch of conditions in our glacial history defines the start of the most recent period of geological time: the 'Quaternary'. It is the Quaternary ice age that has left a clear mark on the landscape. It is the Quaternary glacial advances and retreats that are encoded in sediments in the deep oceans. It is the Quaternary ice age that people are usually referring to when they talk about 'the Ice Age'. The older glacial periods are really the remit of geologists, but once we start to consider the Quaternary things get interesting for the rest of us as well. These are the glaciers of the here and now.

Just as our picture of the world's glaciations over 4,000 million years gets progressively clearer as we approach the present, so the details of the Quaternary become clearer as we look at the most recent parts of its story. The last (present) cycle of glacial activity started around 115,000 years ago, reaching full strength about 75,000 years ago. The oscillations of climate throughout this period seem to be strongly influenced by characteristics of the Earth's orbit, such as the direction and angle of tilt of the Earth's axis and the precise shape of its orbit around the Sun. Periodic variations in these parameters at frequencies of about 26,000, 41,000 and 100,000 years seem to have been a primary control on the pattern of ice ages throughout the Quaternary and are commonly referred to as the 'pacemaker' of the ice ages.

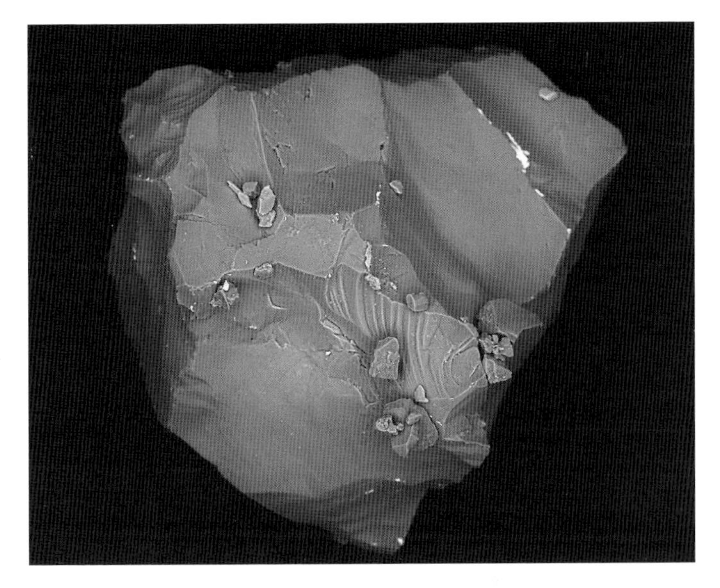

A grain of quartz, about 200 microns (one-fifth of a millimetre) across, from the basal ice layer of Fountain Glacier in Bylot Island, seen through a scanning electron microscope. The grain shows fracture patterns characteristic of subglacial transport.

These astronomical forcing cycles are known as Milankovitch cycles after the scientist who did much to develop the theory. The series of climatic peaks and troughs culminated in a final major glacial peak known as the Last Glacial Maximum about 18,000–25,000 years ago. All of these peaks and troughs in the climate records are given names and labels. Some of the labels, such as the 'Marine Isotope Stages', relate to global markers in the history of ocean or atmosphere chemistry recorded in ice cores or ocean sediments. Other labels relate to more local, recognizable markers in the geologic record or locations to which ice sheets are known to have extended. The glacial period since about 115,000 years ago is referred to as the Devensian in Britain, the Weichselian in Europe and the Wisconsin in North America. The interglacial period that preceded it is called the Ipswichian, the Eemian or the Sangamonian, depending on where you live. It might seem odd to retain local names and systems when there is a global reference available, but the problem is that things don't happen at the same time everywhere. The precise dates of ice advance and retreat in the North American Wisconsin and the European Weichselian do not exactly match up with each other or necessarily with the precise delineations of the Marine

Isotope Stages. As we look closer we see more detail, but the detail does not necessarily make the picture any clearer.

At the Last Glacial Maximum around 20,000 years ago the Greenland and Antarctic ice sheets, which survive today, were larger than they are now, and there were additional ice sheets in other locations including North America and Northern Europe. About one-third of the Earth's land area was covered by ice. The total volume of ice was three times the present-day volume. From about 20,000 years ago the ice started to retreat, albeit marked by periodic readvances. A significant readvance around 11,000 years ago (known variously as the Younger Dryas, the Late Glacial Maximum or the Loch Lomond Stadial, depending on your perspective) was the last hurrah of glaciers in Great Britain. The ice sheet was gone from Britain by about 10,000 years ago, from northern Russia by 9,000 years ago and from Scandinavia by 8,000 years ago. The Cordilleran Ice Sheet over the North American Rockies was gone by about 10,000 years ago. The Laurentide Ice Sheet that covered much of Canada and the northern USA was the largest of them all, but despite some major readvances and late surges it progressively retreated, segmented and was all but gone by about 7,000 years ago. Fragments of the Laurentide Ice Sheet remain: the ice cap on Baffin Island is a

Erratic blocks such as this chalk boulder south of Lyons in France, transported by glaciers and left geologically out of place, provide clues to the history of the ice ages and the former movement of ice.

After long periods of retreat, many glaciers today lie behind substantial moraines or extensive areas of recently deglaciated ground that indicate the glacier's recent former extent. Franz Josef Glacier, New Zealand.

remnant of this ice sheet, and some of the ice buried beneath the ground in Canadian Arctic permafrost has probably survived there as a fossilized relic of it.

From about 10,000 years ago we enter the most recent phase of geological history, 'the Holocene', which, on a short perspective, we might think of as being 'post-glacial' but on a longer scale is just an interstadial within a continuing glacial period. There are still ice sheets in both the northern and southern hemispheres, extensive glaciated regions at high latitudes and mountain glaciers at high altitudes even at the equator. Throughout the last 10,000 years the distribution of glaciers has continued to shift, leaving evidence of advance and retreat in terrestrial glacial landforms and in ocean and lake sediments. Even within historical times we can identify substantial glacier fluctuations. In medieval times conditions were relatively warm and many glaciers retreated. This so-called 'Medieval Warm Period' was when Vikings settled Greenland and started farming, but the warm period was followed by a cold spell lasting several hundred years, known as the 'Little Ice Age'. The Viking settlements did not survive it. Glacial moraines, ridges of debris deposited by advancing glaciers in the Little Ice Age, mark the most recent forward points of many glaciers around the world, from which they have subsequently retreated. Because of the Little Ice Age, the typical image of a glacier today involves a fringing moraine, possibly with a lake dammed up behind it, some marks on the hillside showing former positions of the glacier and some shrunken, sad and retreating ice.

The future is uncertain for these shrinking glaciers, but on the timescale we have been looking at in this section, we still live in a glacial period rather than a non-glacial period. We live in an ice age, but during a relatively non-extreme phase of it. We really do not know if we are in the middle or near the end, and we do not know what will happen next. This story has shown that Earth's current glacial status is part of a long, repetitive, complex and ongoing story.

4 A Short History of Glacier Science

Short histories commonly divide glacier science into the period before 1840, when nobody knew much at all about glaciers, and the period after 1840, when Swiss scientist Louis Agassiz enlightened the world about Earth's glacial story.[1] Longer explanations show that to be an oversimplification, but glacier science certainly emerged into its modern state in the nineteenth century. In the early 1800s little was known about glaciers and it was not recognized that they had once been much more extensive. By 1900, after a century of scientific controversy, most of the basic ideas that underpin modern glacier science were understood, and the Glacial Theory – the idea of a great ice age that covered continents with ice sheets – was widely accepted. It was a scientific revolution – a paradigm shift – of huge significance.

In the 1830s Charles Darwin undertook geological excursions to Cwm Idwal in North Wales and to the so-called 'Parallel Roads' of Glen Roy in Scotland, a series of horizontal terraces that appeared to have been cut into the hillside. Darwin did not consider the possibility that the Welsh landscape had been created by glaciers, and he attributed the Parallel Roads to former high sea levels. Then, in 1840, Darwin heard Louis Agassiz explain his Glacial Theory – the idea that glaciers had once covered many parts of the world that are now free of ice. In fact the landscape of Cwm Idwal was dominated by features created by glaciers, and the Parallel Roads of Glen Roy were shorelines cut into the hillsides by lakes dammed up behind glaciers that had once blocked the valley. Darwin later referred to his previous

Portrait photograph by John Adams Whipple, *c.* 1860, of Louis Agassiz (1807–1873), who did much to bring the idea of 'the Ice Age' to prominence in the 19th century.

conclusion about the Parallel Roads as a 'gigantic blunder', and after revisiting Cwm Idwal and recognizing the evidence of glaciation, admitted that 'a house burnt down by fire did not tell its story more plainly than did this valley.'[2]

Today, glaciers are at the heart of our understanding of climate change and our predictions for the future of life on our planet, but although glacier science has changed in the last two hundred years, some core ideas have remained constant. In 2011 UNESCO published a glossary defining more than four hundred terms used in the study of glaciers, including, of course, the word 'glacier', of which the first recorded use is in French in 1332.[3] The chair of the group that compiled the glossary, Graham Cogley, noted that the earliest use of the word 'glacier' in English, in 1744, had all the ingredients of our twenty-first-century definition, including the recognition that glaciers are not stationary blocks of ice but actually move across the landscape. The first part of the word, *glac-*, comes from the Latin for ice, and the suffix *-ier* suggests the idea of carrying or supplying something, so a glacier is literally a deliverer of ice. That notion of glaciers as moving bodies that transfer ice (and other things) from one location to another is central to the modern understanding of glaciers. Geologists often describe glaciers as conveyor belts for eroded rock debris, and understanding glacier motion and debris transport is at the heart of the study of glacial landscapes.

Early glacier science and belief

In cultures familiar with ice, glaciers featured prominently in traditional histories. Nordic myths saw the universe begin with a clash between the good and evil of fire and ice. Icelandic legend refers to a kingdom of ice, and a primeval Ice Giant whose blood – the melting ice – caused catastrophic flooding. Glacial environments remained mysterious and poorly understood to most people, partly because of their inaccessibility. We invent explanations for the world around us by drawing on our own experiences. People who were unfamiliar with glaciers, but who knew the biblical story of Noah's flood and had experienced actual floods, invoked flooding as the cause of major features in the landscape. In 1650 James Ussher, the archbishop of Armagh, published a chronology of the Old Testament that calculated the date of Earth's creation to be 4004 BC. This gave the Earth an age of less than 6,000 years, during which time the major events that could have altered the landscape were recorded in biblical or historical records. There was a record of Noah's flood, but no biblical record of major ice ages. The so-called diluvial theory, the idea that Noah's flood was responsible for large parts of our modern landscape, was dominant in Earth sciences in the early part of the nineteenth century, partly because history left no time for an ice age.

The Columbia Glacier, Alaska, in 1980. This glacier is the tidewater (sea-terminating) glacier descending from the Chugach Mountains into Prince William Sound.

This was despite day-to-day experience in glaciated areas that clearly demonstrated the work of glaciers. In the 1740s the glacier Nigardsbreen in Norway advanced into cultivated land and destroyed houses. Farmers who lost their land wrote to the king and asked for their taxes to be adjusted in recognition of the loss of land. The local minister recorded that between August 1742 and August 1743 the glacier moved forward 60 m (200 ft), increased in width and demolished houses. To people living in glaciated areas, the propensity of glaciers to advance and retreat, covering different areas of land at different times, was no mystery.

The nineteenth century: a slow (glacial) scientific revolution

Louis Agassiz presented his 'Glacial Theory' to the world in 1840, but the ideas he delivered had been emerging from several independent sets of observations in the preceding decades. It was a revolution in scientific thinking, but it was a long, slow revolution. In the eighteenth century, Swiss engineer Pierre Martell described rocks that had been carried into the valleys by former extensions of Alpine glaciers, and Bernhard Friedrich Kuhn described how former glaciers had altered the landscape in areas now free of ice. In 1795 James Hutton suggested that Alpine glaciers had previously been much more extensive and had spread the granite boulders that were now to be found scattered across limestone bedrock in the Jura Mountains. In 1824 Jens Esmark put forward his own theory of continental glaciation based on the distribution of glacial landforms and deposits. The same ideas were evident in the work of Albrecht Reinhard Bernhardi, a German professor of forestry, who in 1832 proposed that a 'colossal

The Columbia Glacier in Alaska is now an icon of spectacular rapid glacial retreat, but when this photo was taken in 1909 parts of the glacier margin were advancing into a forested landscape.

sea of ice' from the north had been responsible for landforms of the North European Plain. He mentioned the work of other observers including Esmark in Norway as well as Johann Hausmann (1782–1859) and Franz Joseph Hugi (1791–1855) elsewhere in Europe. These were no longer just isolated observations by lone researchers, but a growing body of coherent observation and theorization. German botanist Karl Friedrich Schimper incorporated glaciation into lectures he gave in 1835–6, and coined the term 'Ice Age' (*Eiszeit*) in 1837. In Edinburgh in the 1820s Robert Jameson gave lectures referring to the possibility of a period of extended ice cover. James Forbes attended Jameson's lectures in 1827–8 and his notes indicate that Jameson presented evidence that there had once been glaciers in Scotland. Venetz, Renoir, Perraudin, Charpentier ... the history of glacier science includes a long list of names of people who had recognized and reported evidence before 1840 that glaciers had once been much more extensive and had left behind a signature in the landscape. However, these scientific discussions remained outside the mainstream and were not embraced by the general scientific community, let alone by popular consciousness. Then, in 1840, one man brought all these ideas together into a persuasive story and promoted it vigorously.

Louis Agassiz argued that only the theory of former widespread glaciation could explain many features observed in the landscape, and in 1840 he published his two-volume work *Études sur les glaciers* (Studies on Glaciers). In the same year he visited Glasgow for the meeting of the British Association for the Advancement of Science. During that visit he toured Scotland with British geologists and they encountered convincing, widespread evidence of former glaciation. Suddenly the idea that had been growing steadily for fifty years was properly launched into the scientific mainstream. It was in 1840 that glaciers and the ice age became big news. We noticed glaciers, seriously, for the first time and the scientific turn towards glaciers as major parts of our global system started. After Agassiz, Charles Darwin adopted the theory. James Croll was inspired by it in his development of the theory of astronomical forcing of climate change. The idea

A satirical portrait of the geologist William Buckland, by T. Sopwith. Buckland is shown carrying maps of ancient glaciers and walking on ground marked by 'prodigious glacial scratches'. One pebble is labelled as being scratched by a glacier before the Creation, and another scratched by a cart wheel the day before yesterday.

rattled through the sciences and was adopted as core knowledge in science and in culture more widely.

Acceptance of the Glacier Theory was by no means immediate or unopposed. For decades the debate continued about the relative significance of glaciers as opposed to the traditional mechanisms of icebergs and floods in altering the landscape. One focus of controversy was the scattering across the landscape of 'erratic' boulders that were out of place geologically: granite boulders mysteriously located in areas of limestone bedrock, for example. In the last years of the nineteenth century the Glacialists' Association had its origin in a meeting calling itself the 'North-West of England Boulder Committee', formed to organize the systematic investigation of these erratics. The first edition of the *Glacialists' Magazine* (1893) noted, 'The geologists of Hull have organised a "Boulder Committee" similar to that appointed by the Yorkshire Naturalists' Union, which has done such valuable work in the past few years.'[4] In a string of articles between 1839 and 1855 Charles Darwin attributed glacial boulders in various locations to the rafting of boulders on floating icebergs. Darwin's work included observations of boulders on

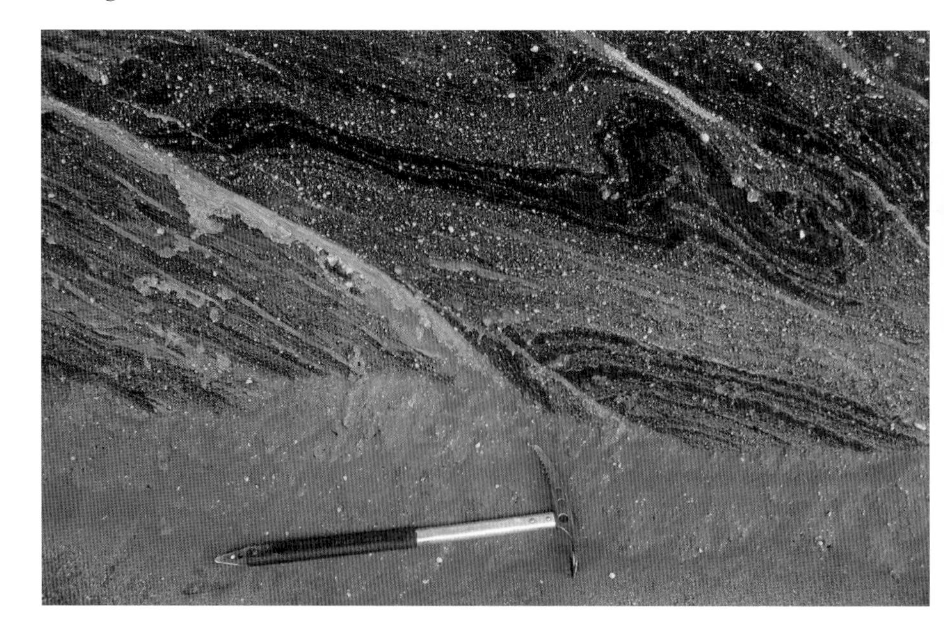

This close-up view of ice and debris close to the base of the glacier Skeiðarárjökull in Iceland illustrates the kinds of deformation structures within the basal ice layer that allow glaciologists to work out how the ice behaves and how glaciers interact with the ground beneath them.

Tierra del Fuego during the voyage of the HMS *Beagle* in 1833, and it was as recently as 2009 that modern glaciologists finally published evidence to contradict Darwin's iceberg theory and demonstrate that 'Darwin's Boulders' in Tierra del Fuego had actually been transported and deposited by glaciers.[5] Some debates take a long time to settle. Another big controversy in the nineteenth century was the notion that ice could ever erode solid rock to excavate valleys. According to naturalist John Ruskin in a speech to the Royal Institution in 1863, 'A glacier scoops out nothing . . . and can no more deepen its receptacle than a custard can deepen a pie-dish.'[6] Even at the end of the century, T. G. Bonney in 1893 wrote in the *Geographical Journal* that glaciers were not very effective agents of erosion, certainly not compared with 'torrents', and in his 1893 book *The Story of Our Planet* he wrote: 'A glacier . . . excavates only under exceptional circumstances and to a limited extent.'[7] Some scientists struggled to shed the old idea of the Great Flood.

One obstacle to progress was the lack of scientific theory to explain puzzles such as the climate changes that would allow glaciers to expand or retreat, but this theoretical gap started to close in the second half of the nineteenth century. In 1864 James Croll produced a theory of ice ages based on astronomical calculations of periodic variations in the Earth's orbit around the Sun. James Geikie used Croll's work in synthesizing data from around the world and in 1874 Geikie's landmark book *The Great Ice Age* was published with the declared intention of showing how the deposits previously attributed to floods or icebergs were in fact outcomes of glaciation. The Glacial Theory became firmly established and the rate of scientific discovery about glacial processes, environments and landforms accelerated. In his report on the 1899 Harriman expedition to Alaska, the geologist G. K. Gilbert wrote in 1903: 'The growth of knowledge of Alaska glaciers is so rapid that a summary of existing knowledge would have but transient value.'[8]

Throughout the nineteenth century and into the twentieth, scientific discoveries about glaciers were closely related to expeditions of geographical discovery and exploration, if only because

so many of the world's glaciers are in areas that were still largely unexplored. The South Polar expeditions of the early twentieth century are often described primarily in the context of the 'race for the pole', but J. W. Gregory, the scientific advisor to the British National Antarctic 'Discovery' expedition of 1901–4 (before he resigned in a dispute regarding the priority of science over other aspects of the expedition), set out specifically glaciological goals

At the margin of Margerie Glacier, Alaska, debris that has been eroded and transported by the ice is exposed and released as the ice melts or calves into Glacier Bay. The

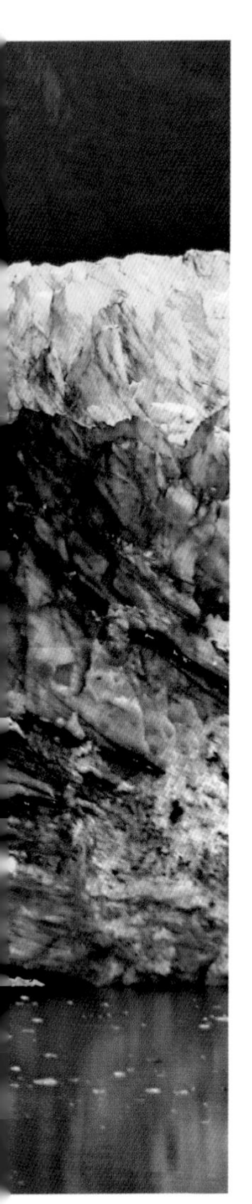

debris provides a wealth of information about subglacial environments and landforming processes.

for the expedition, including study of ice crystallography to help explain glacier motion. The 'Antarctic Manual for the Use of the Expedition of 1901' included instructions for making glaciological observations, and Gregory specifically related the study of modern glaciers to the goal of understanding how past ice ages affected the landscape of Europe.

The idea that 1840 marked a sharp, clear break between ignorance and enlightenment is clearly oversimplified – science rarely works that way. Revolutions take time, and there are always counter-revolutionaries. Nevertheless, by the end of the nineteenth century many aspects of glacier science were starting to look very much as they do now.

The twentieth century: details, paradigm shifts and reinventions

A glaciologist from the start of the twentieth century would find many things familiar in the science at the start of the twenty-first. In 1994 the glaciologist Louis Lliboutry pointed out that all of the processes of glacial erosion that were known to modern glaciologists had already been identified by 1900.[9] Writing in 1888, W. L. Rogers made the assertion that the biggest issue in the understanding of glaciers was figuring out exactly how gravity and heat control glacier flow.[10] A century later, reviewing the state of the discipline, Richard Alley called that elusive 'flow law' for glaciers the holy grail of glacier science. However, that continuity is only part of the picture.[11]

The emergence of new technologies has been a driving force behind developments in glaciology. One innovation that had a huge impact was the development of photography. In his 1903 report of the 1899 Harriman expedition to Alaska, G. K. Gilbert wrote: 'For the study of changes in the size of glaciers photographic views are of peculiar value.' Photography was the first of many technological enhancements to our visions of the ice. In 2012 the American Geophysical Union hosted a conference session on 'Enhanced Glaciological Understanding through Innovation'. Glaciologists now routinely apply remote sensing,

This 1907 photograph by the Norwegian Institute for Svalbard and Ice Sea Exploration, shows the explorer Adolf Hoel (right), who was later the founding director of the Norwegian Polar Institute, at Lilliehöök Glacier in Kings Bay, Svalbard.

LiDAR, photogrammetry, time-lapse photography, GPS, seismic monitoring, marine geophysics and so on to the monitoring, recording and visualization of glaciers and glacial processes.

More than facilitating clearer observation, modern techniques have opened up whole avenues of research that were not possible in the nineteenth century. For example ice-coring technology, which was confined to shallow drilling in the time of Agassiz, now allows us to extend deep cores through the 4,000-m (2.5-mi.) thickness of the Antarctic Ice Sheet. New dating techniques allow us to assign ages to different layers in a glacier, correlate those with records of atmospheric composition preserved in the ice, and reconstruct detailed histories of environmental change. Cosmogenic nuclide analysis has permitted dating of glacial landscapes themselves, so that we increasingly have well-dated reconstructions of glacier extent and landscape evolution. Geophysical techniques now allow fast and cost-effective sub-surface investigations of glaciers and glacial sediments. Many glaciologists today use satellite data, ground-penetrating radar and mass spectrometers where in former times they might have used ice axes, ropes and shovels.

The development of these techniques has been driven by, and has facilitated, new types of questions. Glaciologists increasingly turn their attention to links between the glacial system and the oceans, atmosphere and human systems in a changing climate. Nevertheless, the history of progress has in some ways been stuttering and hesitant, constantly revisiting old questions with the benefit of new techniques and long experience. Sometimes, even in modern science, knowledge is lost or forgotten. The study of the history of science should help prevent us reinventing forgotten wheels, but in glaciology there are fascinating examples of science repeating itself. Discoveries have been made, written up . . . and then forgotten, only for the same questions to be asked again a generation later.

In 2001 the Proceedings of the Geologists' Association included a paper that the authors described as 'the first comprehensive description and interpretation of Pleistocene glacigenic deposits exposed in a cliff section at Thurstaston on the Wirral

A transport flight delivering scientists high onto the surface of the Greenland Ice Sheet to work on the North Greenland Eemian Ice Drilling (NEEM) ice-coring project in the summer of 2009.

Peninsula, NW England'.[12] It was a splendid paper, providing a comprehensive description and analysis. It even referred to previous literature about the glaciation of the UK from as early as 1860 to set the historical context. However, it omitted to refer to one particular publication from that era: an obscure note published in the *Glacialists' Magazine* in 1895, which was itself a detailed study of those same sediments.[13] The 1895 paper seems to have fallen into the great pit of forgotten knowledge.

A century apart, both papers set out to address the same questions about the glacial deposits referred to variously as 'till', 'diamicton' or 'boulder clay', and the 1895 research had used much

Chris Fogwill holds an ice core that has been extracted with a KOVACS ice corer from ancient ice exposed close to the surface of the ice sheet in the Ellsworth Mountains, Antarctica.

A section of the cliff of glacial till at Thurstaston, Wirral, that has been discussed in detail by researchers from different generations. Dog included for scale.

the same evidence as the 2001 team. The list below compares extracts from the 2001 publication with extracts from the note published a hundred years earlier:

> 2001: The sedimentary succession at Thurstaston is best explained by the advance and subsequent recession of a single terrestrially based ice sheet . . . the diamictons are interpreted as basal, deformation tills with the interbeds of the sand, gravel and mud lithofacies as indicators of subglacial meltwater flow.

> 1895: The boulder clays I attribute, then, to the direct action of an ice sheet, and the gravels and sands to subglacial drainage.

> 2001: There is no evidence at Thurstaston to suggest a glacio-marine origin for the Late Devensian deglaciation sediments on this margin of the Irish Sea basin.

> 1895: I cannot see any evidence in the deposits on this shore in favour of the view that the beds were deposited in deep water.
>
> 2001: [There is a] lack of clearly defined changes in clast lithology through the various lithofacies . . . Particularly distinctive rocks include the Borrowdale Volcanic Group tuffs, the Ennerdale granophyre and Eskdale granite.
>
> 1895: Both beds contain shell fragments and their boulder contents appear to be similar. The latter consists of Eskdale granite, Buttermere granophyres, Scotch granites, Lake District volcanic rocks.

It is reassuring to know that science can repeat its results in an independent restudy of the same question after a gap of a hundred years. But it is astonishing that Victorian scientists over a century ago could reach the same conclusions as scientists in the twenty-first century, only to have their work lost and forgotten.

One big difference between the 1895 and 2001 stories of the Thurstaston till was that in 2001 the tills were clearly identified as 'deformation tills', which are created when sediment

A glacially striated erratic emerging from the till cliff at Thurstaston, Wirral, indicating the provenance of the till and hence the direction of flow of the ice that deposited it.

underneath an ice sheet is deformed by the ice above. This mechanism of glacier motion and till formation was not mentioned in 1895 because at that time nobody knew that glaciers moved in that way. This idea did not emerge until the 1970s, and was heralded as a paradigm shift in glaciology. Glaciologists previously assumed that glaciers moved by sliding across the ground underneath them or by deforming internally. Once it had been suggested, it seemed obvious that large areas of present and former ice sheets rested not on rigid bedrock but on soft sediments. In the decades that followed, the idea of subglacial deformation was applied to the solution of several major glaciological and geomorphological problems, including the mechanisms behind the extraordinary fast flow of ice streams, the explanation of periodic ice-sheet surges, and the origin of that most enigmatic of landforms, the drumlin.

So, while many of the basics of glacier science were established by the start of the twentieth century, that century saw more than a simple consolidation of knowledge; it witnessed some substantial technological, theoretical and practical advances. But as that century closed, there were still big issues or 'grand challenges' facing the discipline.

Big issues for the twenty-first century

Glacier scientists have some clear points of focus in the early twenty-first century. Reviewing the state of the subject in 2006, geographer David Sugden wrote, 'A report card on the progress of glacial studies over the last 40 years or so might read "encouraging progress, but surprisingly large gaps in knowledge remain".'[14] Sugden listed some grand challenges, including the challenge of interdisciplinary collaboration, limitations to the resolution of our dating techniques, the theoretical and empirical foundations of ice-sheet modelling, and understanding large components of the glacier system such as the North Atlantic ocean circulation and the West Antarctic Ice Sheet. Glaciologists in the twenty-first century are asking big questions, and need both detailed data and sophisticated theory to approach answers.

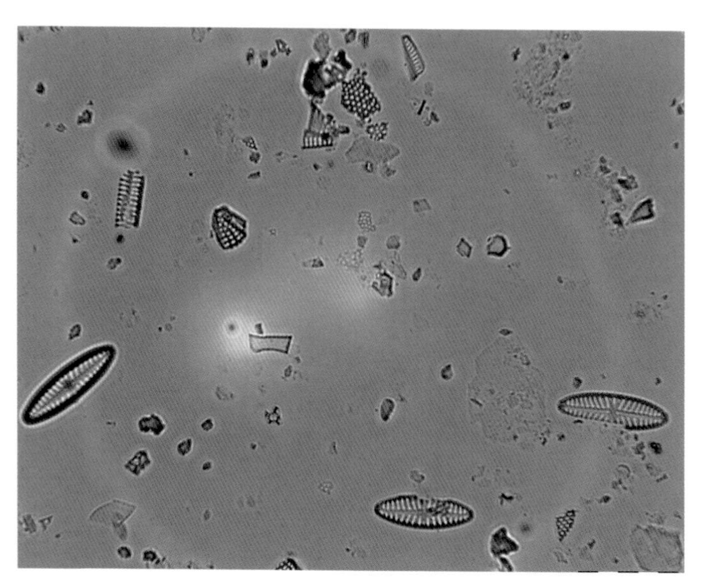

Because living diatoms have specific environmental tolerances, diatoms in the fossil record, for example in lake or ocean sediments, provide a record of how environments have changed in the past. This microscope image shows diatoms just a few hundredths of a millimetre long.

Some of our big questions are much like those being asked 150 years ago: where did glaciers extend to at different points in history? How did those former glacial episodes relate to climate change? What can we say about former glaciers on the basis of the landscapes they left behind? But today's glaciologists also look to the future. How might we predict and measure the response of glaciers to climate change? And still we need to focus on the mechanisms of glacier motion, mass balance and fluctuations in order to predict how glaciers will behave in the future. What would be the effect of increasing water flow at the base of the Antarctic Ice Sheet? What will happen to ice-sheet velocity if floating ice shelves disappear? Could sections of the Antarctic Ice Sheet collapse, with catastrophic impacts on global sea level? Answering these questions requires a fundamental, detailed understanding of glacier mechanics, along with detailed monitoring of glacier behaviour. We need our ablation stakes as well as our numerical models – even if some of the work of metaphorical ablation stakes might now be accomplished by measurements made remotely from airborne or satellite missions.

There are still many things about glaciers that we simply do not know. Even their basic geography is elusive. Glaciers are hard to measure either on the ground or from space, and they are constantly changing. The year 2012 saw the publication of 'The First Complete Glacier Inventory for the Whole of Greenland'.[15] This was based on satellite data mostly acquired between 1999 and 2002. Glacier inventories are out of date even before they are published. The glaciological community recognizes this as a major issue, and the 'Global Land Ice Measurements from Space' (GLIMS) initiative tries to maintain a database of current glacier extents. In association with the International Panel on Climate Change (IPCC) reporting process, the Randolph Glacier Inventory offers a 'Globally Complete Inventory of Glacier Outlines'.[16] It still faces problems of detail and in being kept current, but is a major step forward, and indicates the importance of this basic data. It is often assumed that satellite data provide a clear

A DHC-3 Otter involved in surveys of Alaskan glaciers as part of NASA's Operation IceBridge, which was designed to cover the data-gap in polar observation between the decommissioning of the ICESat-1 satellite in 2010 and the launch of ICESat-2 in 2018.

and detailed picture of exactly where everything is on the Earth's surface. For glaciers, hard to see clearly beneath snow and cloud in the mountains, constantly shifting position through the seasons and over the years, this is not the case. As for what glaciers will do in the future, we are again at a loss. Some models suggest that once glaciers reduce below a certain level they will not return even if climate recovers to pre-industrial conditions, whereas other models suggest that glacier shrinkage is reversible if climate change is halted. We really do not know.

We remain part of a scientific tradition closely connected to the work of our predecessors. Glacier science has traditionally been dominated by 'big science', because major logistics and significant funding are required for large-scale glaciological research. Organizations such as the International Association of Cryospheric Sciences and operations such as NASA's Ice, Cloud, and Land Elevation Satellite (ICESat) mission point to the continuing institutionalization of glaciological research. In a polite euphemism for the debates that characterize research, scientists often refer to their disciplines as 'plural and contested'. What they mean is that different scientists have different ideas, and that they continue to argue about them. Glaciology is populated by mathematicians, geologists, geographers, physicists and

On 18 June 2018, at Vandenberg Air Force Base in California, technicians work on the Advanced Topographic Laser Altimeter System (ATLAS) on NASA's Ice, Cloud and Land Elevation Satellite-2 (ICESat-2) prior to its planned launch.

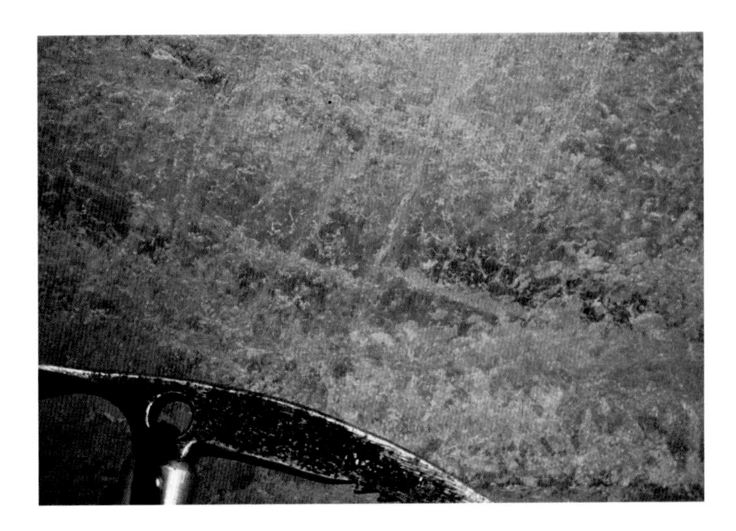

The chemical composition of air bubbles trapped within the ice, and of the ice itself, can tell us a lot about the composition of the atmosphere in the past when the ice first formed from snow, and about the processes that have affected it since.

many others. It is little wonder that they sometimes struggle to maintain a common point of view. Nevertheless, one fundamental scientific tradition that we continue is the tradition of reading what has previously been unread from the story embodied in the ice and in the landscape. In 1857 Henry Wadsworth Longfellow wrote a poem to celebrate the fiftieth birthday of Louis Agassiz that included these lines pointing to that same tradition:

> And Nature, the old nurse, took
> The child upon her knee,
> Saying: Here is a story-book
> Thy Father has written for thee.
> Come, wander with me, she said,
> Into regions yet untrod;
> And read what is still unread
> In the manuscripts of God.

DIAGRAM OF

METEOROLOGY,

DISPLAYING THE VARIOUS PHENOMENA OF THE ATMOSPHERE.

Drawn and Engraved by John Emslie. Published by J. Reynolds 174 Strand, Sept 20th 1846.

REFERENCE.

1 *Effects of Tempestous winds on land*
2 *Effects of do at Sea. The Malstrom.*
3 *Waterspouts.*
4 *Fog.*
5 *Clouds Stratus.*
6 *do Cumulus*
7 *Clouds Cirrus.*
8 *do Nimbus or rain cloud.*
9 *do Cirro Cumulus.*
10 *Rain.*
11 *Snow.*
12 *Perpetual Snow*
13 *Glaciers*
14 *Aurora Borealis*
15 *Rainbow*
16 *Halo.*
17 *Mirage*
18 *Parhelia or Mock Suns*
19 *Zodiacal light.*
20 *Ignis Fatuus or Will with a Wisp.*
21 *Lightning.*
22 *Lightning Conductor.*
23 *Falling Stars.*
24 *Aerolites.*

8

5 Glaciers and the Big Global System

Glaciers are part of a complex, interconnected global environmental system. Their existence depends on a particular set of climatic conditions and is controlled by atmospheric circulation, the hydrological cycle, the positions of the wandering continents, the heights of mountain ranges, the tilt of the Earth's axis, the luminosity of the Sun and many other factors. Changes in any one of these will affect the occurrence, properties and behaviour of glaciers. And they are all changing, all the time.

What makes glaciers advance and retreat?

The most direct way in which glaciers reflect environmental change is their expansion and contraction at timescales ranging from days to millions of years. These fluctuations reflect the amount of ice being supplied to the glacier by processes of accumulation (such as snowfall) and the amount being lost by processes of ablation (such as melting). This 'mass balance' is largely related to climate. If more ice is supplied than is lost the glacier will grow, and vice versa. The edge of the glacier is the position where the supply of ice from up-glacier is finally extinguished by ablation as the ice moves forward.

From 1846, an engraving by John Emslie showing glaciers and perpetual snow as part of what we would now call the global climate system.

Even when there is no long-term change, many glacier margins oscillate backwards and forwards a little every year in response to seasonal changes in ablation close to the margin. During summer, melting outweighs the forward movement of the ice and the margin retreats. During winter, ablation is low

Two Landsat satellite images of the Upsala Glacier, draining from the eastern side of the South Patagonian Icefield. The first (left) was acquired in January 1985; the second (right) was acquired in February 2017. In the intervening 32 years the glacier margin retreated by about 10 km (6¼ mi.).

and the forward movement of ice exceeds the rate of ablation, so the margin advances. Over longer periods, changes in ablation and accumulation can be reflected in more substantial episodes of advance or retreat. For fluctuations related to changes in accumulation, there can be a substantial time lag between the change in accumulation high up on the glacier and the response at the margin, as the effect of the change is slowly fed down through the glacier. By contrast, changes caused by local ablation at the margin take effect immediately. Therefore, changes in the position of a glacier margin represent the combined effects of recent changes in ablation close to the margin and historical changes in the rates of both accumulation and ablation further up-glacier. The advance or retreat that we see is therefore hard to associate directly with a specific individual climate event. We see a composite response.

Climate is not the only control on glacier fluctuations. Sometimes, different glaciers that are very close together and in the same climatic regime can behave very differently as they respond to different forces. For example, in surge-type glaciers, periodic variations in friction at the bed of the glacier lead to sudden changes in the speed of ice movement. When the ice moves faster, it can travel further into the ablation zone before it melts, so the margin of the glacier advances. This advance has nothing to do with climate but is controlled entirely by the glacier's internal dynamic processes. Glaciers with margins that float in water, rather than resting on land, also behave differently. Land-terminating glaciers respond to changes in ice supply by advancing or retreating so that the ablation area grows or shrinks until ice loss balances ice supply. Floating glaciers, for example in a fjord, lose mass via the calving of icebergs, and their ability to do that depends not on their extent but on the depth of the water they float in and the width of the fjord at the margin. So once a floating glacier starts advancing or retreating it will not stop until it reaches a position where the shape of the fjord is just right for the glacier to calve icebergs at the appropriate rate. Even a small climate-related perturbation nudging the glacier

The glacier Briksdalsbreen, Norway, 1869.

The glacier Briksdalsbreen, Norway, 2013.

margin forwards or backwards could result in an advance or retreat that is disproportional to the climate change. It is risky to make inferences about climate change on the basis of the retreat or advance of floating glaciers.

Even in a warming world where most glaciers are retreating, the complexity of the system is such that some glaciers in specific situations may advance in spite of the general global trend. To be able to read glaciers as ledgers of environmental processes we need to be aware of some of the complexity in the relationship between glaciers and the global environment.

The Brückner and Heim Glaciers flow into Johan Petersen Fjord in southeastern Greenland.

Feedback and complexity

Glaciers are controlled by climate, ocean circulation, atmospheric composition and many other environmental factors, but they also have an impact on them in return. It is typical of the environmental system that things affect each other in two-way relationships, and that any change in one component of the system sets off feedbacks that rebound onto whatever made the change. For example, a change in global climate leading to the expansion of ice sheets would lead to an increase in the overall reflectivity (albedo) of the Earth's surface. In turn, that increased albedo would cause more of the Sun's energy to be reflected (rather than absorbed) by the Earth, making the climate even cooler. Glaciers would then expand more, albedo would further increase, and the temperature could continue to fall in a 'positive feedback', amplifying the original cooling. Conversely, warming of high latitudes by ocean currents such as the Gulf Stream might cause increased melting of ice in Greenland, but the release of that cold, low-salinity water into the North Atlantic could affect the ocean circulation so as to reduce the impact of the northward-flowing warm currents, inhibiting the high-latitude warming in a 'negative feedback' that would curtail the initial melting.

These multi-directional relationships between climate, ocean circulation and glaciers, with their complexities and feedbacks, make predictions difficult and reveal the intricacies of a natural system that remains difficult to describe, let alone explain with confidence. This is illustrated by the following extract from a report by scientists modelling the way a climate phenomenon known as the North Atlantic Oscillation (NAO) affected the mass balance of glaciers in different places:

> In southwestern Scandinavia, winter precipitation causes a correlation of mass balances with the NAO. In northern Scandinavia, temperature anomalies outside the core winter season cause an anti-correlation between NAO and mass balances. In the western Alps, both temperature and winter precipitation anomalies lead to a weak anti-correlation of

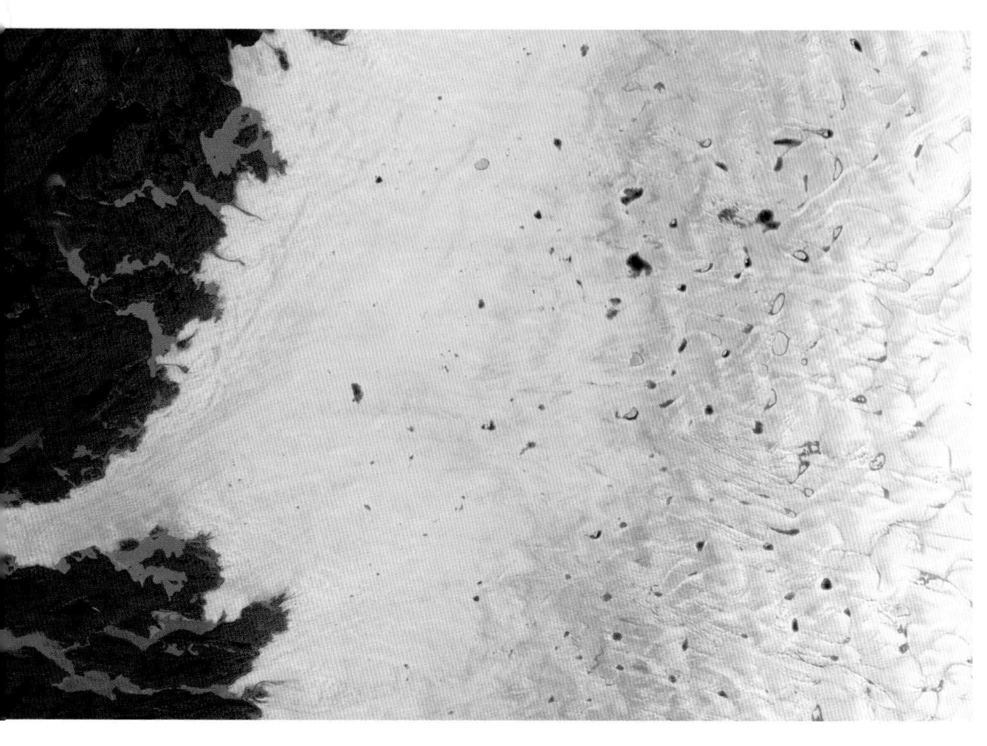

Landsat satellite image showing a section of the western part of the Greenland Ice Sheet, 15 July 2015. The coast of Greenland is on the left of the image, and the ice-sheet surface is marked by meltwater lakes.

mass balances with the NAO, while in the eastern Alps, the influences of winter precipitation and temperature anomalies tend to cancel each other, and only on the southern side a slight anti-correlation of mass balances with the NAO prevails.[1]

Glaciers and sea level

One of the environmental relationships most under scrutiny in the context of future environmental change is the relationship between glaciers and sea level. The growth and decay of glaciers – especially the world's great ice sheets – are closely connected to rises and falls in global sea level. Glaciers are part of the global hydrological cycle. Water evaporates from the oceans and from the land, travels in the atmosphere, returns to the surface mainly as rain or snow, and makes its way overland or through groundwater back towards the oceans. Glaciers are part of the transport system between the atmosphere and the oceans, and a location for the storage of water. Roughly speaking, on a global average, water that travels through the system via rivers takes about ten days to make its way back to the sea. Water that follows a route going through a glacier typically takes, on average, about 10,000 years. Glaciers take water temporarily out of the cycle and delay its return to the sea. When glaciers expand and large ice sheets exist on the continents a lot of water is being held back from the oceans. When those ice sheets melt, a lot of water is returned to the oceans.

At present, about 26 million cubic kilometres of water are locked up in glaciers, the vast majority of it in the Antarctic Ice Sheet. That is equivalent to a depth of about 65 m (213 ft) of water spread across the world's oceans, and is one of the reasons that people are worried about climate change and the melting of ice sheets. There is a lot of ice, and as it melts it will release a lot of water. At the height of the last ice age about 20,000 years ago, when there were ice sheets over North America and Europe, the ice locked up in glaciers was equivalent to 197 m of water spread

A small meltwater stream descending from the glacier-capped Kharkhiraa mountains in northwest Mongolia.

across the world's oceans, and the sea level was more than 10 m (33 ft) lower than today. At the end of the ice age about two-thirds of the ice melted back into the oceans. As glaciers retreated gradually over thousands of years the rise in sea level was also gradual, estimated at about 40–50 mm (1½–2 in.) per year. However, there were occasional jumps. For example, the sudden unleashing into the Atlantic Ocean of water from the Agassiz-Ojibway lake system that had been ponded up behind the retreating Laurentide Ice Sheet in North America led to abrupt sea-level changes of several tens of centimetres. A single major release of meltwater about 8,200 years ago is thought to have had such an impact on ocean conditions in the North Atlantic that it slowed down the ocean circulation and led to a widespread climatic cooling event. These sea-level rises, and the associated impacts on the see-sawing postglacial climate, would have been felt by humans. These events seem long ago but they are within the era of human occupancy of the Earth.

The idea that sea-level change is caused by the growth and decay of ice sheets seems quite simple. The volume of water in the oceans changes, so the level of the oceans changes. However, as always, there are complicating factors. One of the most important in this context is isostasy. The weight of several thousand metres of ice resting on the Earth's crust causes the crust to sink a little. Sections of the crust under different weights are depressed by different amounts. Land beneath 3,000 m (2 mi.) of ice will be depressed, slowly, by about 1,000 m (3,300 ft). When the ice is removed, the crust will slowly rise back. Geologists often refer to the crust 'bouncing' back, but since the uplift occurs at about the same rate as the growth of fingernails, and since parts of the Earth's surface are still rising a few millimetres each year in response to the unloading of ice at the end of the last ice age, 'bounce' might be too exuberant a term.

At the end of the last ice age coastlines globally were inundated by rising seas as glaciers melted and released their stored water. Subsequently, however, the gradual isostatic rebound of land masses where there had been thick ice caused localized falls in sea level as the land rose. The history of sea-level change

Satellite imagery has been an important technological innovation for the study of glaciers. This image from NASA's *Terra* satellite in August 2014 clearly shows Iceland's ice caps, with bare ice showing in their ablation areas and snow covering their accumulation zones.

is therefore geographically and historically very complicated. Because different parts of the same coastline can rebound by different amounts and at different rates, former coastal features that were originally all at the same height can be tilted. Raised beaches around the west coast of Scotland, for example, have been raised less in the far west than in areas close to the thick centre of the former ice sheet. In Finland, the southern coast is rising at about 2 mm per year, whereas areas further north are rising at as much as 8 mm per year. As the whole southern part of Finland is gradually being tilted to the south, water is being slowly decanted from lakes and released into the Gulf of Finland. The amount of uplift in different areas depends on how

This image from the Space Shuttle *Columbia* shows many different glaciers in the St Elias Mountain Range in southern Alaska, including the Malaspina Glacier, which is the largest piedmont glacier in the world.

much depression previously occurred, which is itself controlled by the thickness of the ice mass. We can therefore use the distribution of rates of uplift to reconstruct the shape and thickness of former ice sheets.

Isostatic depression and uplift also affect the ocean floor. A depth of water of 100 m (328 ft) being added to an ocean basin by ice melt would gradually depress the ocean floor by several tens of metres, so the initial rise in sea level caused by the influx of water would slowly be partly offset by depression of the sea floor. Similarly, in reverse, when water is taken out of the oceans and locked up in ice-age glaciers, the initial drop in water level will gradually be partially compensated by long-term isostatic rebound of the sea floor.

Ocean currents, atmospheric composition, and climate

Pouring water in such huge volumes from melting glaciers into the sea affects not only the level of the water but its composition and the way it flows around the planet. Ocean currents are driven by water temperature and salinity in a system called the 'thermohaline circulation' (*thermo* for temperature, *haline* for salinity). The temperature and salinity of water affect its density, making it rise or sink within the ocean. The thermohaline circulation is sometimes described as a global conveyor belt carrying water around the planet. Surface currents and currents deep beneath the surface often flow in opposite directions, transferring energy around the planet as water is heated at the equator and cooled closer to the poles. In high latitudes, warm ocean currents travelling polewards from the equator close to the ocean surface deliver energy to the atmosphere as they cool. As the water cools it sinks and returns towards the equator at depth. These currents have a major impact on the climate. One important part of the circulatory system is the 'North Atlantic meridional overturning circulation'. The label 'meridional' indicates that this part of the ocean system operates on a broadly north–south axis between equatorial and polar regions. The 'overturning' part of its name refers to the way that warm, northward-flowing surface water cools, sinks and flows back south at depth. This 'North Atlantic Deep Water' plays a major part in driving the whole global system, and is interesting to scientists because it seems that it can be switched on and off. Changing the amount of cold, low-salinity water that is poured into the North Atlantic from ice sheets such as the Laurentide Ice Sheet (when it was still there) and the Greenland Ice Sheet (which still is) can have a big impact. Some people predict that increased meltwater coming off the Greenland Ice Sheet in the near future could disrupt the circulation, switch off the Gulf Stream and lead to severe cooling of the North Atlantic. This would have serious impacts on the climate of the east coast of North America and the west coast of Europe. Other people have suggested that the process is already happening, and that weakening of the Gulf Stream explains why the North Atlantic is the only part of the world defying the overall trend of global warming.

Glaciers can also affect climate in other, less direct ways. For example, changes over time in the amount of chemical weathering of rocks at the Earth's surface can lead to changes in the amount of CO_2 in the atmosphere, and the amount of weathering is strongly related to the advance and retreat of glaciers. When glaciers cover a lot of the ground, rock weathering is inhibited, especially if the glaciers are frozen to their beds. But as glaciers retreat they typically leave behind a lot of broken-up rock fragments, silt and clay, and they often generate a lot of water. Just as a lump of sugar will dissolve more quickly in coffee if it is crushed up into smaller grains to expose more surface area to the hot liquid, so rock will weather more quickly if it is broken up into fragments exposing more surface area to meltwater and the elements. When ice sheets retreat, a lot of fresh weathering occurs on the newly exposed land, and weathering of rocks typically involves chemical reactions that draw CO_2 from the atmosphere. Increased weathering as the major ice sheets retreated about 10,000 years ago could have been responsible for the marked reduction in atmospheric CO_2 that has been observed in ice-core records from that time.

Glaciers can also affect atmospheric CO_2 through their impact on volcanic eruptions. Extensive glaciation is thought to inhibit volcanic activity, and even when volcanic activity does occur beneath thick ice sheets there is less chance of volcanic gases reaching the atmosphere. However, as ice sheets thin and decay, the release of pressure on the crust can unleash volcanic activity, which in turn can increase the production of atmospheric gases and aerosols that affect climate. Those climate changes then impact once more on the expansion and retreat of glaciers.

Ancient environments recorded in ice cores

Many of the connections between glaciers and the rest of the global system are recorded in the ice as it builds up, and can be

Sediment being transported from the glacier system into the ocean. Sam Ford Fjord, on the northeast coast of Baffin Island, Nunavut, Canada.

reconstructed from ice cores drilled through glaciers. Louis Agassiz, one of the pioneers of glacier science, drilled cores up to 60 m (197 ft) deep in Switzerland in the 1840s. As technology improved, successive efforts achieved progressively deeper cores and by the 1960s expeditions to Greenland and Antarctica were extracting cores of ice thousands of metres long, reaching all the way through the ice sheets from the surface to the bed. The deeper into the glacier the core penetrates, the older the ice it accesses. The ice is mainly formed from snow that fell on the surface and was gradually buried. As the snow was buried, so too were pockets of air between the snowflakes, and the particles of dust that fell with them from the atmosphere. Each layer of snow carried slowly down into the glacier takes a time capsule of snow, air and dust down into the ice. Glaciologist Richard Alley has written a book about ice coring called *The Two-mile Time Machine*, and that title describes exactly what an ice core is.[2] Looking at the bottom part of the longest cores we see ice

The drilling trench, excavated into the surface snow high on the Greenland Ice Sheet, from which the NEEM deep-drilling operation was conducted to extract an ice core that penetrated into ice dating from the last interglacial period.

that fell as snow hundreds of thousands of years ago, and the ice core gives us layer by layer a running history of atmospheric composition all the way back into prehistory. Ash from the volcanic eruption of the Icelandic volcano Laki in 1783 has left a prominent marker horizon buried in the ice that has been found in cores from Greenland and the Canadian arctic. Measurements of lead content in ice cored from the Quelccaya Ice Cap in Peru show lead pollution from silver smelting by the pre-Columbian Incas. The changing levels of lead through the core show that pollution increased in the sixteenth century after the Spaniards increased the scale of mining, tailed off in the nineteenth century as the economy languished, but soared again in the twentieth century as new mines were opened and as cars started to burn leaded fuel. Lead that went into the atmosphere from Roman smelting 2,000 years ago can be found at depth in Greenland ice cores, and shows that the Romans caused as much as a fivefold increase in lead levels in the atmosphere. Nearer to the surface, the Industrial Revolution shows up clearly in the cores. Radioactivity soars in the mid-twentieth century as a record of nuclear weapons testing. The Chernobyl disaster of 1986 has a clear horizon of its own.

A glacier is like an environmental repository, or a global memory. Connected to everything, it recalls information about everything. Ancient ice in a present-day glacier tells us how the planet has changed through the glacier's lifespan. Geological evidence of where glaciers previously existed tells us about how climate has changed over long periods of history. How glaciers are presently behaving tells us about what is happening right now in the big interconnected global environmental system. There are many reasons to keep an eye on glaciers: they are the canary in the coal mine for environmental change, but if they suffer the canary's fate their disappearance will have further implications for our environmental future.

6 Glacier Economics: Hazards, Resources, Politics

Only a small minority of the world's population lives in the immediate vicinity of a glacier, so the idea that glaciers play a major role in our day-to-day lives and economy may seem improbable. However, even those of us who do not live close to glaciers are deeply affected by them. We all live with global atmosphere, climate and ocean systems strongly influenced by glaciers. For their water supply huge numbers of people, especially in South Asia, rely on rivers that are fed by glaciers. Many of us live in towns built on glacial sediments, eat food grown on glacial soils or take holidays in spectacular glacial landscapes. Road and rail networks follow lines in the landscape that reflect the passage of glaciers or glacial meltwater tens of thousands of years ago. Low-lying islands in the tropical oceans as far from glaciers as it is possible to be are experiencing catastrophic consequences as sea levels rise in response to the melting of distant ice. Even for those of us who consider glaciers to be far away or long ago, their presence is significant in our daily lives.

Some glaciers have become more significant as they have become more accessible. For example, the precise positions of some high-altitude political boundaries in glaciated areas such as the Himalayas and the European Alps were essentially notional when nobody ventured much into the mountains. Now, with satellite surveillance, increased technological potential for accessing resources in remote areas, and growing international tensions over territory and resources, some previously little-known glaciers are becoming strategically significant. The Siachen Glacier on India's disputed border with Pakistan has become the site of a long-running conflict that has claimed the lives of more than a thousand soldiers. Disputes between Italy, France and Austria remain unresolved despite high-quality mapping, partly because glaciers themselves change position and extent over time. Glaciers, constantly shifting, do not make for stable political boundaries. When the mummified 3,000-year-old body of Ötzi the Iceman was discovered in glacier ice between Italy and Austria, the two countries could not agree about which borders the body was discovered within. As water resources become critical, and as more countries introduce legislation about their glaciers, border disputes over glaciers are increasingly likely to arise. In 2016 the Indian government drafted a law that could impose fines of up to U.S.$15 million on any cartographer or organization that misrepresented India's borders. As those borders in some glacial areas are disputed, this is potentially problematic for international mapping organizations such as Google. In 2010 Argentina was among the first to put glacier protection into law, with its 'Regime of Minimum Standards for the Preservation of Glaciers and Periglacial Environments'. Other nations with glacier protection law include Chile, Columbia, Ecuador,

President Obama at the Exit Glacier, Alaska, in 2015, collecting meltwater from the front of the glacier.

On the front line between Austrian and Italian forces, the Marmolada Glacier in the Italian Dolomites played a key strategic role during the First World War, and troops built a substantial network of defences and accommodation in tunnels within the ice.

Pakistan, Kyrgyzstan, Switzerland and Austria. In 2016 Jennifer Cox wrote about the legal status of glaciers in Canada and concluded that they were essentially unprotected under existing environmental law.[1] Glaciers need legal protection from not only the indirect effects of climate change but the direct destructive impact of extractive industry. Glaciers have become an important part of our social as well as our physical systems. At the Kumtor Mine in Kyrgyzstan, extraction of gold from beneath the Lysii and Davidov Glaciers has involved the excavation of a pit into the ice and the dumping of rock waste onto the glacier surface. The work has led to changes in the glacier flow regime and demonstrated how, in the search for mineral resources, glaciers might be completely destroyed by mining operations.

In some parts of the world, people have lived close to glaciers throughout history and experience the direct effects of glaciation in very obvious ways. Describing the relationship between people and glaciers in South America, historian Mark Carey writes, 'Peruvians have suffered the wrath of melting glaciers like no other society of earth.'[2] In Peru 25,000 people have died in glacier disasters in the Cordillera Blanca since 1941. In Europe, the death toll is lower but glacier hazards are widespread and significant. In 2003 the European Community GLACIORISK project identified 206 'risky' glaciers in Europe where ice avalanches, glacier floods or other dangerous events had occurred. They identified 721 recorded deaths from glacier hazards, including more than four hundred deaths in glacier floods and more than two hundred deaths in ice avalanches.

Glaciers affect people in good ways as well as bad. Whether something is a hazard or a resource depends on your point of view. Glacier meltwater is a problem for the bridge builder but a boon for the hydropower engineer. The extent to which hazards can be mitigated, or resources exploited, depends partly on our level of technological development. Over time, something that was once perceived as a hazard can become a resource and vice versa. Events that were once of little interest can become significant as the development of remote areas increases human exposure to glacier hazards. One of the largest periodically draining ice-dammed lakes in North America, Strandline Lake in Alaska, used to drain through an area of uninhabited wilderness. That area now contains oil and gas installations, roads, bridges and power lines leading to Anchorage, Alaska's largest city. The development of glaciers as a resource for adventure tourism has brought increasing numbers of people into contact with glacial environments and their associated risks. About 600,000 tourists visit New Zealand's Fox Glacier and Franz Josef Glacier every year, and following a series of fatalities local government has recently considered whether steps should be taken to restrict or close public access to the terminal face of the glaciers. We are starting to take glaciers more seriously.

Preparing for the worst in a hazardous environment, Pararescuemen from the U.S. Air Force 304th Rescue Squadron practise their survival and rescue techniques among the glaciers of Mount Hood, Oregon.

Hazards

Some hazards, such as crevasses treacherously concealed by snow, are confined exclusively to the area of the glacier itself. Others, such as meltwater floods, ice avalanches and debris flows extend their reach well beyond glacier limits. All of them change position over time, as the margins of the glaciers advance and retreat. Glacier fluctuations can overwhelm land and destroy settlements, threaten water supplies, divert river systems and cause disruption of transport routes. In 1986 the front of the Hubbard Glacier in Alaska advanced and blocked the mouth of the tributary Russell Fjord. With its exit to the sea blocked, the Russell Fjord was turned into an ice-dammed lake. It started to increase its depth and as it filled with fresh water its salinity changed, threatening the sea life trapped in the fjord. There was also a threat that the filling lake would overspill its bank to the south, flooding into the Situk River and threatening Yakutat Airport. In fact, before the lake overspilled, the ice dam at the mouth of the fjord gave way, releasing more than 5 cubic kilometres of water into the sea over a period of 24 hours. The flow was equivalent to about 35 times the flow of water across the Niagara Falls, and was one of the biggest glacier flood discharges observed in modern history. A similar blockage occurred again in 2002 and will happen again in the future if there are repeated advances of the Hubbard Glacier or its tributaries.

During the Little Ice Age, a period of lower temperatures affecting most of the globe between the thirteenth and nineteenth centuries, substantial glacier advance was a feature of life in many glaciated areas. At Breiðamerkurjökull in southern Iceland, which is now largely uninhabited with ice reaching close to the sea, records from the early settlement of Iceland (AD 870–930) through to the eighteenth century describe a landscape of

Icebergs in a periodically draining ice-dammed lake at the margin of the Greenland Ice Sheet.

Ngozumpa Glacier, Nepal. The glacier surface shows signs of significant melting, and lakes are developing both on the ice surface and along the glacier margin.

woodland and settlements disappearing beneath advancing ice. The settlement of Fjall, thriving in 1660, was abandoned by 1695 and was under the ice by 1709. In the French Alps, at the start of the seventeenth century glaciers in the Mont Blanc area encroached on substantial areas of land, destroying farms and villages. There were petitions to the tax authorities for exemptions based on loss of land. Two villages involved in one such petition, Châtellard and Bonanay, were eventually engulfed by the advancing ice. There is now a tourist trail through the locations of the lost villages.

The nature of a glacier hazard can change through time. In 1678 the residents of Fiesch and Fieschertal in Switzerland approached Pope Innocent XI for his blessing for a prayer against the growth of the Aletsch Glacier. In 1862 the villagers started to hold annual processions to reinforce their prayer. But as the Little Ice Age drew to a close, advance turned into retreat and environmental change saw widespread decline in mountain glaciation. By 2009 Fiesch and Fieschertal decided to revoke their prayer against the glacier. Over three centuries it had retreated 3.5 km (2.1 mi.) and, if anything, prayers were needed now to help the glacier, to support tourism and safeguard the water supply. In the annual procession in 2012 crucial words in the liturgy were changed and the prayer was reversed. The prayer now calls for the salvation of the glacier: 'Ice is water, water is life.' In 2015 the pope delivered an encyclical, an official statement, on the dangers of climate change and the Vatican held a conference under the title 'Protect the Earth, Dignify Humanity'.

The most devastating glacial disasters have been ice avalanches where ice at the front of a glacier on a steep slope breaks away, or 'calves', and plunges down hill. The largest-known ice avalanche, from Mount Iliamna, Alaska, in 1980, involved about 20 million cubic metres of ice. In 1597 an entire village near the Simplon Pass was buried beneath an icefall, and in 1965, 10 million cubic metres of ice fell from the Allalingletscher in Switzerland, killing 88 people working on the construction of a hydroelectric power plant in the valley below. In designing reservoirs that are likely to be in the path of ice avalanches, care must be taken to ensure that the dam is sufficient to accommodate waves caused by ice falling into the water. In China, floods from ice-dammed and moraine-dammed lakes have been described as the single most important glacier hazard.

Two of the worst ice avalanches in history occurred in the same location – Nevado Huascarán, the highest Peak in Peru's Cordillera Blanca. In 1962, 4,000 people were killed when 10 million cubic metres of ice fell 3,000 m (9,800 ft) from the glacier known as Glaciar 511 into the populated valley below. The avalanche travelled 16 km (10 mi.) in eight minutes, burying several villages and the town of Ranrahirca in icy rubble up to 20 m (66 ft) thick. Bodies were found as far as 160 km (100 mi.) downstream at the mouth of the Santa River on the Pacific coast.

In 1970 a rock and ice avalanche from the same mountain buried several villages and most of the town of Yungay, killing more than 20,000 people.

One of the most deadly glacier-related disasters was the lahar (volcanic mudflow) caused by melting of snow and ice during the eruption of the volcano Nevado del Ruiz in Colombia in 1985. About 10 per cent of the volcano's ice cap melted, and meltwater combined with volcanic debris to create lahars which flowed down channels off the volcano. Peak flow in one channel reached 48,000 cubic metres of debris per second, with a velocity of 38 km per hour (23½ mph) and a wave front about 40 m (131 ft) high. The lahar flooded through the town of Armero at about 30 km per hour (19 mph), killing more than 20,000 people. The lahar had been predicted, and the threat to Armero was well understood. The terrible death toll has been attributed to the unwillingness of government authorities to risk the economic and political cost of premature evacuation.

Many volcanoes are capped with glaciers that pose this kind of threat. The eruption of Mount Redoubt, Alaska, between 1966 and 1968 removed about 60 million cubic metres of glacier ice from the upper part of the Drift Glacier, causing a series of meltwater floods, one of which flooded the site of the oil-tanker terminal on Cook Inlet at the mouth of the Drift River. The same volcano erupted again between 1989 and 1990, and more than 100 million cubic metres of ice were dislodged, causing floods that required evacuation of the oil terminal.

Glacial meltwater floods are often referred to by the Icelandic term *jökulhlaup*, which means a leap in water level caused by the glacier. A jökulhlaup from the Icelandic glacier Mýrdalsjökull in 1918 had a discharge equivalent to three times that of the Amazon. In southern Iceland on 5 November 1996, meltwater from a subglacial volcanic eruption burst from the glacier Skeiðarárjökull. Discharges of around 45,000 cubic metres per second swept away the roadway and bridges from a section of Iceland's main national ring road. The Icelandic prime minister announced that four hours of flooding had knocked back the road-building programme by thirty years.

This 1911 photograph of a railway line on the Kenai Peninsula, Alaska, shows the effects of a meltwater flood from the Spencer Glacier.

Disastrous floods afflicting the Cordillera Blanca in Peru started in the 1930s after a period of about two hundred years without floods. The flooding began because glaciers started to retreat in the 1920s, and lakes formed between retreating glaciers and their moraines. As the lakes filled, the moraines became unstable and liable to collapse either because of the pressure of water or because the water level rose to a point where waves caused by ice falls into the lake could overtop the dam. In 1941 a flood killed 5,000 people and devastated the town of Huaraz. Immediately afterwards, work was started to mitigate the hazard of the Cordillera Blanca lakes by the installation of artificial drains and canals through the moraines. These restricted lake depths to prevent dam failure. Engineering solutions such as these to reduce the magnitude or likelihood of glacier-related flooding have a long history. The Giétroz Glacier in Switzerland advanced from a tributary valley to block the main valley of the Val de Bagnes, creating an ice-dammed lake in 1549, 1595 and 1818. In 1595 the lake overflowed and killed five hundred people. When the lake formed again in 1818, engineers cut an artificial channel across the dam, and drained about one-third of the water before the lake burst. Nevertheless, fifty people perished in the resulting

Engineers in 1911 at the snout of the Spencer Glacier on the Kenai Peninsula in Alaska, building a dam to divert the glacier's meltwater stream.

flood. In 1892, a subglacial water pocket of more than 100,000 cubic metres broke out from the Tête Rousse Glacier, killing 175 people. Engineers proceeded to drill a tunnel through rock to the bed of the glacier to drain other possible water pockets in order to prevent further casualties. In 2016 engineers in Nepal embarked on work to lower the level of Lake Imja near Mount Everest as it threatened to break through its moraine dam. At about 5,000 m (3.1 mi.) above sea level it is one of the highest disaster mitigation projects in the world.

None of the floods recorded during historical times matches the scale of floods that occurred at the end of the ice age when the great ice sheets were melting. The best known of these are the Missoula floods in North America and the Altai floods in Central Asia. The Missoula floods occurred when water ponded up against the edge of the Laurentide Ice Sheet was repeatedly released in catastrophic outbursts that spilled across North America, carving the spectacular landscapes of the channelled scablands. The Altai floods were caused by the cataclysmic release of water from lakes that were dammed up behind glaciers descending from the Altai Mountains. The Missoula and Altai floods are the largest-known floods of fresh water ever to have occurred, and nothing equivalent to their magnitude has happened during recorded human history.

Glacier hazards are not confined to the glaciers themselves. Icebergs are not glaciers, but fragments that have broken off or 'calved' from a glacier into water. Icebergs pose a threat to shipping, oil platforms and seabed installations such as pipes and cables. Up to 2,500 icebergs each year drift southwards along 'iceberg alley' from Greenland towards the Grand Banks area east of Newfoundland. In that area alone, between 1882 and 1890, 54 ocean liners were reported sunk or damaged in collisions with icebergs. The most famous victim was the RMS *Titanic*, sunk with the loss of more than 1,500 lives in 1912. After that event, the International Iceberg Patrol was established to monitor icebergs in the North Atlantic. Attempts to destroy icebergs by bombing and other means have largely been unsuccessful, but towing them for short distances to divert them from collision with installations such as oil rigs has become an established practice. The company Atlantic Towing Limited describes itself as a leader in the field of 'ice management'. More informally, the work of diverting icebergs is referred to as 'iceberg wrangling'. The first iceberg-proof fixed oil platform, the Hibernia, was established in the Grand Banks off Newfoundland in 1997. The Hibernia was designed to withstand the impact of icebergs up to 6 million tonnes, the largest size of berg expected to be able to reach it in the 80 m (262 ft) of water in which it stands.

Resources

Glaciers have a long history of threatening human activity, but also a long history as a resource. One of the chief threats posed by glaciers in a changing environmental future is the potential loss of glacial resources on which we rely.

One of the earliest uses of glacier ice was as a refrigerant. Caves within glaciers have been used as stores for perishable provisions, and ice cut from glaciers has been used commercially and domestically for preserving food and cooling drinks in areas far removed from naturally occurring ice. Before refrigerators were invented, Norway exported ice to other European countries, and ice houses for the kitchens of the European aristocracy were stocked from distant glaciers. In the 1850s, ice was transported from Alaska to California, and small icebergs from southern Chile were transported as far north as Peru for use in refrigeration. The traditional market in glacier ice continues to the present day, albeit on a tiny and local scale, in areas such as the Cordillera Blanca of Peru and in the Hunza region of the Karakoram Mountains. However, these historical uses of glaciers pale into insignificance against their role in other aspects of our global economy.

Glaciers are an abundant source of the most basic of human resources – water. Meltwater from glaciers is the source of many of the world's major rivers, supports agricultural populations in some of the most densely populated parts of the planet and provides water to a significant number of major cities. Many arid areas including the Thar Desert and the Atacama Desert receive water for irrigation from glaciers in adjacent mountains. Himalayan glaciers are major contributors to river systems such as the Indus and Ganges. Glaciers are especially valuable as a water source since they produce most water in hot dry weather, when other sources are at a minimum. In the Bolivian capital La Paz, about a quarter of the dry-season water supply is provided by glacier meltwater. The city of Boulder, Colorado, derives about 40 per cent of its water from the Arapaho Glacier catchment and the Silver Lake reservoir that it feeds. The progressive loss of glacier-fed water supplies is a serious issue.

Icebergs, being composed of fresh water, would offer a potential water resource if they could be transported to areas where water is needed. In the 1970s Prince Mohammed al Faisal of Saudi Arabia established Iceberg Transport International and galvanized the scientific community to consider issues related to iceberg utilization. Faisal's ultimate goal was to transport a 100 million-ton iceberg about 14,500 km (9,000 mi.) from Antarctica to Saudi Arabia. Problems associated with transporting and processing the icebergs were such that the idea did not come to fruition in the twentieth century, but in the face of global environmental change that threatens the water security of many wealthy parts of the world there are still organizations planning to exploit icebergs as resources both for their water and for the energy that can be released by the melting of the ice.

More realistic glacier-power opportunities are already being exploited. Most important of these is the use of glacier meltwater to generate hydroelectricity. This kind of power has a particular advantage over other renewable sources such as solar and wind energy in that the raw material – summer meltwater – can be stored in reservoirs until the winter when demand for power is greatest. In places such as Greenland, energy requirements could theoretically be met entirely by hydroelectric power from glacier meltwater. In Switzerland, glacier-fed hydropower accounts for more than 50 per cent of the country's electricity. In one Swiss scheme, the meltwater stream from the Oberaar

An Air Greenland helicopter approaching the Black Angel mine portal approximately 600 m (1,970 ft) above the Qaamarujuk Fjord in central West Greenland. The view across the fjord shows an unnamed glacier descending from the ice cap on the Alfred Wegener Halvø Peninsula.

One of the primary headstreams of the River Ganges close to its starting point at the snout of the Gangotri Glacier.

Glacier, at the head of the Aare River, was dammed in the 1930s to create an artificial lake that has been used ever since as part of an increasingly complex system of engineered waterworks. Comprising seven hydro-dams and 130 km (80 mi.) of water pipeline, this one hydro network supplies 7 per cent of Switzerland's electricity.

At the other end of the energy-production cycle, it has been suggested that glaciers could be used as natural dustbins for the long-term storage of radioactive industrial wastes. These wastes need to be kept out of contact with the biosphere for periods of up to 250,000 years. In the early 1970s, the International Atomic Energy Agency considered the possibility of burying waste beneath an ice sheet. One plan involved radioactive waste being fused into glass and loaded into lead-shielded containers for burial within the ice. Critics of the plans pointed out huge potential dangers. The stored material would need to be sufficiently difficult to retrieve that no potential malefactor would be able to gain access, but not so difficult that the material could not be retrieved if necessary. A method would be needed to prevent the waste containers from melting through the ice and reaching the glacier bed where they might be damaged and leak. Research in the 1980s concluded that none of the proposed methods was safe. Since then, developments both in international security and in the pattern of global environmental change have reinforced that conclusion.

Many formerly glaciated areas are rich in glacier-related sand and gravel deposits that are a major industrial resource. In Britain alone the building industry uses around 50 million tonnes of non-marine sand and gravel each year, and deposits from glacial meltwater comprise a substantial part of that material. The distribution of these materials reflects the geography of former glaciers. For example, fluvioglacial landforms such as eskers, kames and outwash fans are easily identifiable sand and gravel sources.

These glacial deposits form the ground on which many human landscapes are built. Glacial deposits are immensely variable, even over very short distances, because of the variety of materials and the range of different processes by which they are deposited. Deposits related to specific glacial processes have specific engineering properties. For example, lodgement tills are likely to be highly consolidated, clayey and unsorted, and typically are difficult to excavate but have good bearing and stability characteristics. By contrast, fluvioglacial materials are likely to be less consolidated and hence easier to excavate but less stable. In many parts of the world, an understanding of the geotechnical properties of glacial deposits is an important part of any construction or environmental engineering project.

In the last few hundred years glaciers have been a source of aesthetic interest to travellers. This has increased in the last fifty years with the growth of skiing as a popular holiday activity and the development of large-scale adventure tourism. Both the USA and Canada have 'glacier' national parks. The tourist industries of Greenland and Iceland are in large measure based on their glacial landscapes. National parks around the world are based either on present-day glaciers (for example Glacier Bay National Park in Alaska) or on landscapes left behind by ancient glaciers (such as the Lake District National Park in the UK). More than half a million people each year visit New Zealand's Fox Glacier and Franz Josef Glacier. In some cases glaciers become a destination for spiritual rather than purely aesthetic reasons. The Gangotri Glacier at the source of the Ganges has become such an attraction to tourists and pilgrims that human activity has been blamed for accelerating the retreat of the glacier. Such proximity between glaciers and large numbers of visitors brings risk, but for local economies it can bring economic opportunity.

In the Earth's uncertain future, environmental change is likely to alter the ways in which glaciers and people interact. As a period of globally retreating glaciers is established, new areas are exposed to glacier hazards. For example, the history of problems associated with glacier retreat and the formation of hazardous meltwater lakes in Peru is starting to repeat itself in new locations such as Bhutan and Nepal in the Himalayas. The pattern of glacier resources will also change. In the Rockies and the Andes, glaciers that supply city reservoirs are diminishing

towards extinction, and in the Alps the melting ice that supplies the reservoirs that feed the power stations is becoming more and more distant. But as the threats and opportunities change, so do the technologies with which we approach them and the global economic framework within which we do so. A future is in prospect that includes both deliberate interventions in glacier systems by humans, and more extreme consequences for economies and societies as glaciers change.

Engineering and construction in glacial settings typically involves working in extreme environments. This image shows construction of the Glacier Moosfluh ski lift in Switzerland.

CAT

7 Glaciers in Art

The artist Wilhelmina Barns-Graham visited Switzerland's Grindelwald Glacier in 1948 and produced a series of pictures in which she recalled specifically 'the massive strength and size of the glaciers, the fantastic shapes, the contrast of solidarity and transparency . . . in a few days a thinness could become a hole, a hole a cut out shape losing a side . . . It seemed to breathe!'[1] Barns-Graham's reaction included elements common to many artistic observers of glaciers: a sense of glaciers being vast, grand and overwhelming but at the same time fragile and ephemeral, combined with an impression of their being somehow almost alive.

Glaciers move people and they serve as metaphors, but both the metaphor and the way it moves us have changed through time. To the eighteenth-century Romantics, glaciers were the awful and sublime. To twenty-first-century eco-artists they are endangered victims. But certain motifs remain constant across time and between media. A glacier is huge and inexorable, yet ephemeral. It is impenetrable and ancient, but in a state of perpetual change. It moves at a rate that puts it in a timeframe outside human consciousness. It reminds us how small we are, and how short-lived. Persisting into our present from a period before there were people, made of snowflakes that fell before any human impact on the planet, glaciers are symbols of lost purity. A landscape re-emerging from beneath a retreating ice sheet is given a fresh start. Glaciation is the ultimate clean slate. Containing in their ice and their air bubbles the last surviving relics of that clean past, but melting away before our eyes, glaciers are emblems of environmental fragility and the human threat to the planet.

Barns-Graham was not the first European artist to be inspired by the massive, fantastic, almost animate characteristics of glaciers. As the Romantic movement emerged in the eighteenth century, glaciers featured prominently in the imposing landscapes of artists who visited the wild country of the European Alps. Some, such as Carl Hackert in his hand-coloured etching of around 1780, *Vue de la source de l'Arveron*, treated the subject in sufficient clarity and detail that the exact positions of the glacier margin and the geological characteristics of specific rocks can be identified from the pictures. For others, such as Philip de Loutherbourg in his 1803 painting *An Avalanche in the Alps*, the glacial landscape is not a precise record of geological detail but a backdrop to a human drama being played out in the foreground. In de Loutherbourg's painting *Travellers Attacked by Banditti* (1781) the backdrop of ice-capped peaks contributes to the

Carl Hackert, *Vue de la source de l'Arveron*, *c.* 1781, hand-coloured etching on paper. Prior to the Romantic period many paintings of glaciers were so precise that they have been used to make historical reconstructions.

William Pars, *The Rhône Glacier and the Source of the Rhône*, *c.* 1770, watercolour and Indian ink. This image provides a quite precise representation of the scene.

impression of the physical and social wildness of the Alpine pass where civilized people risk their lives at the hands of savage nature and savage bandits. The scene would work less well in a setting of rolling farmland, or even in mountain country without glaciers. The glaciers provide the savage, Romantic icing on the Alpine cake.

This was a recurrent theme in the works of Romantic artists such as J.M.W. Turner and John Ruskin. Some of Ruskin's glacier art, such as his sketch of the Aiguilles de Chamonix (1849), was in the older tradition of realistic, visually accurate representation, commensurate with his background in architecture and natural history. Partly as a result of studying Turner's work, Ruskin's approach changed and he increasingly used paintings to explore and record an emotional response to landscape, reflected in glacier paintings such as his 1863 watercolour *Mer de Glace – Moonlight*.

In 2006, artist George Rowlett embarked on a project to follow in Ruskin's footsteps with a painting expedition to Chamonix. Rowlett's work, and its popularity, indicates the survival of a long-standing Romantic tradition in which art represents not only the facts of a scene but the feelings. Rowlett's colours, technique and style are distinctively twenty-first century, but his mission to capture the aesthetic and emotional impact of an encounter with the glaciers is the same as Turner's or Ruskin's.

In the nineteenth and twentieth centuries, European styles of landscape painting were exported to European colonies and territories. This was partly due to the emigration of European artists such as Albert Bierstadt (1830–1902), Franz Biberstein (1850–1930) and John Fery (1859–1934) to the New World. In North America in the mid-1800s the Hudson River School and the Rocky Mountain School were inspired and motivated by the European Romantic tradition of landscape painting, and presented mountainous wilderness as a backdrop to comfortable foregrounds designed to indicate the success of human frontier settlement. Whereas European Romantics had presented the Alps as a threatening wilderness, artists such as Bierstadt presented the Rocky Mountains as a wilderness pushed back by the

John Ruskin, *The Aiguille Baltiere*, *c.* 1856, drawing with wash.

progress of civilizing settlement. Wilderness was becoming scenery: the sublime without the terror. The paintings were often on huge canvasses and were exhibited to help encourage political or social goals such as westward expansion or the formation of national parks. John Fery moved to Wyoming in the 1880s in response to a growing demand for pictures of the Western landscape. He painted prolifically in Yellowstone, Grand Teton, and especially in Glacier National Park. He was subsequently commissioned to make a series of large-scale paintings for the Great Northern Railroad, depicting locations along their route to advertise their services and decorate their stations. It was an early step on the road towards glacier as advertising icon. 'See America First' was the slogan that accompanied the images, and Fery's landscapes continued the trend of placing spectacular, awe-inspiring mountain and glacier landscapes as a backdrop to a civilized foreground. In the railroad advertisement paintings

Thomas Fearnley, *Grindelwald Glacier*, *c.* 1837, oil on canvas.

Hermann Herzog (1832–1932), *Fisherman on a Mountain Lake*, oil on canvas. Herzog was one of a significant number of European artists who emigrated to North America and exported their style of painting with them.

the mountains and glaciers were often positioned safely on the outside of a carriage or hotel window, viewed from the comfort of the passenger's seat.

Similar stories can be told for other locations. For example, it has been argued that John Gully's (1819–1888) watercolours of New Zealand instilled a sense of national identity in the colonials who had his paintings on their walls, and 'helped to establish in the minds of newly-settled gentlefolk from Victorian England a taste and love for the wild exotic beauty of New Zealand'.[2] Gully's pictures show huge landscapes with few signs of human occupants. Colonial painting could validate the taking of a land from its former occupants by representing the land as having been empty prior to colonial settlement.

By the early twentieth century, some North American artists were establishing new styles to break away from the European tradition and reflect more local, nationalist priorities. The

Canadian 'Group of Seven' was a school that grew in the 1920s around artists such as Lawren Harris and A. Y. Jackson, committed both to a nationalist art agenda and to representing the unique character of the Canadian landscape. Their legacy is controversial. Critic Michael Valpy denounced the group's 'fraudulent mythology' as perpetuating the falsehood of unspoilt Canadian wilderness despite its long, economically motivated environmental exploitation.[3] Their art shaped how Canadians saw their own country, but critics argue that it encouraged a false vision. Unfortunately, in one critical case, the vision – or at least its official geographical interpretation – was spectacularly flawed. In 1936 the National Gallery of Canada purchased a painting by Lawren Harris entitled *Greenland Mountains*. In an ironic twist, given the nationalist sentiments of the artist and the purpose for which this painting was later used, the painting was incorrectly labelled and put into storage with the title *Bylot Island*, which is in the Canadian Arctic. When in 1967 Canada issued a set of postage stamps based on Canadian landscapes painted by Canadian artists, this painting, mistaken for a picture of Canada's Bylot Island when in fact it was a picture of a site in Danish Greenland, was printed on the Canadian 15-cent stamp. Between 1967 and 1973, Canadian postage stamps mistakenly carried a painting of Danish Greenlandic glaciers.

Art-science links: great expeditions

In the era before photography, scientific expeditions required artistically skilled scientific recorders, if not dedicated expedition artists, and a substantial amount of glacier art derives from the records of polar and mountain exploration. In 1899 the American railroad tycoon Edward H. Harriman set up an expedition of artists and scientists to explore the coastline of Alaska. The team included eminent geologist G. K. Gilbert, who was twice president of the Geological Society of America, and the naturalist John Muir, who was founder of the Sierra Club and one of the leading figures behind the establishment of the U.S. National Parks. Expedition artists included landscape painters Frederick

Dellenbaugh and R. Swain Gifford. Dellenbaugh had previously served as both expedition artist and assistant map-maker on John Wesley Powell's 1871 expedition down the Colorado River. Exploration, art and science comingled. The 1899 Harriman expedition also engaged a photographer, Edward Curtis, who captured over 5,000 images during the expedition. In his introduction to the scientific report, Gilbert wrote that the expedition had 'carried many cameras and secured a large number of views of glaciers'.[4] The same has probably been true of every expedition to glaciated regions since that date.

Perhaps the most famous early glacier photographs come from the 1910–13 Terra Nova South Polar expedition of Robert Scott. This was the first Antarctic expedition to include a professional photographer, Herbert Ponting, and Ponting's official title was 'camera artist'. The chief scientific officer on the expedition, Edward Wilson, was also an accomplished artist. Wilson's paintings documented the expedition's scientific work but also recorded the polar landscape. His landscape paintings and sketches are very precisely drawn, and are typically annotated or titled with the precise location, date and time in a style that betrays the scientist in him, but at the same time many of them evoke the atmosphere and emotional resonance of the vast polar wilderness. For example *Cave Berg off Cape Evans, with Mount Erebus at 5.30 p.m. on September 1, 1911* delivers a clear impression of the fractured texture of the ice, while *Down a Crevasse* gives scant attention to the textures of ice but paints a vivid picture of urgent human drama on the glacier.

Twentieth-century photographic technology did away with the requirement to include artists among an expedition's scientific personnel, but the value of including artists in scientific expeditions is still recognized. For several years until 2009 the British Antarctic Survey hosted an artists and writers programme that included at various times a film-maker, a poet, a glass artist, several landscape painters and a musical composer. In 2005, Argentina's national Antarctic organization established artists in residence as part of an art and culture project to record the work of scientists on the Argentine Antarctic bases. Many of these initiatives have been primarily a cultural outreach on the part of essentially scientific organizations, but others have stemmed more from artistic roots. Prominent among these is the series of Cape Farewell expeditions initiated by artist David Buckland in 2001 to promote a cultural as well as scientific response to the threat of climate change. The clearly stated climate-change framework of Cape Farewell illustrates how the focus of artists changes to reflect changes in scientists' preoccupations. Victorian artists such as John Brett (1831–1902) were enthralled by scientific debates surrounding the existence of ice ages. In 1856 Brett visited the Swiss Alps, and in the following years exhibited a number of glacier paintings at the Tate Gallery and the Royal Academy in London. Brett's painting *The Glacier of Rosenlaui* shows sufficient detail of individual rocks that the painting could have served almost as a geological dataset contributing to the debate that was raging about erratic boulders and the former extent of Alpine glaciers.[5] The tradition of artists being inspired by contemporary scientific issues continues with Cape Farewell, but whereas the Victorian Romantics were elaborating upon and exploring the newly discovered concepts of deep time and ice ages, artists in the twenty-first century are exploring newly discovered potential futures of loss and destruction.

Today, the art in which glaciers feature most prominently is art inspired by environmental concern. Some of this is by artists directly motivated by issues surrounding climate change and the melting of glaciers, and some of it derives from scientists choosing to communicate their messages through art rather than the more confined channels of science. Science and environmental concern in combination often add up to politics, and a great deal of recent glacier art has been the art of environmental politics, often sponsored by environmental political organizations.

Artist John Quigley travelled with a scientific expedition on the Greenpeace icebreaker *Arctic Sunrise* in 2011 to create a huge reproduction of Leonardo da Vinci's *Vitruvian Man* on melting sea ice in the Fram Strait. Greenpeace announced that the piece was constructed 'as a call for urgent action on climate change'.

Rudolf Reschreiter (1868–1939) was a Munich-based artist known for his hyper-realistic depictions of nature. This work, dating from the early years of the 20th century, depicts part of the Hollental Glacier in southern Germany.

Greenpeace expedition leader Frida Bengtsson said the work was intended to highlight the dramatic changes taking place in the Arctic, and to illustrate 'how our dependency on fossil fuels is tipping the balance of the relationship between nature and humans'. In 2010 a global art show on climate change was organized by the '350 Earth' advocacy group, which takes its name from the number of parts per million that is often given as the maximum acceptable level of CO_2 in the atmosphere. The plan was to highlight the hazards of global warming by creating artworks around the planet that could be seen from space.

Mia Baila, *Getting Warmer*, 2013, acrylic on canvas. In this painting, the artist included warm tones in her palette to indicate a warming climate and environmental change.

One of these was a giant image of a polar bear marked out using red emergency tents at the foot of a receding glacier in Iceland. Greenpeace Switzerland developed a glacier art campaign to draw attention to the impact of global warming on Switzerland's shrinking glaciers. Part of the campaign was a photo shoot by New York artist Spencer Tunick involving six hundred volunteers removing their clothes and standing naked on the Aletsch Glacier.

Many present-day artists are aligning themselves with scientific expeditions to document glacial environments, and the 'artist's statements' in their catalogues and websites make clear associations between their art, the science they observe and their environmental fears. According to artist Jill Pelto, art is a uniquely articulate lens through which to address environmental concerns and inspire people to take action. Pelto has produced a series of watercolours in which scientific data such as rates of glacier recession are incorporated as the shape of a mountain or the curve of an ocean wave. Maria Coryell-Martin describes herself as an 'expeditionary artist' in a tradition of artists as naturalists and educators. She paints polar and glaciated regions vulnerable to climate change where she has often collaborated with scientific teams. Her aim: to build a portfolio to share and cultivate environmental awareness. This specific use of art to promote environmental awareness and concern is a common theme among recent glacier artists. Diane Tuft describes her work as 'documenting the beauty and fragility of our ever-changing environment' by capturing images in the ultraviolet and infrared light spectrum, because global warming and ozone depletion are increasing the amount of radiation in those wavelengths reaching the Earth's surface. Tatiana Iliina's 'Glaciers Gone' series of paintings begun in 2007 was specifically intended to raise awareness of the disappearance of the world's glaciers. The notion of loss is echoed widely across the field. An exhibition of paintings and photographs of Rocky Mountain glaciers to mark the 100th birthday celebration of Glacier National Park was titled 'Losing a Legacy'. *WIRED* magazine in 2012 featured 'Last Chance to See: A Photographic Tour of Earth's Doomed Ice'. The article referred to Danish photographer Klaus Thymann, who formed a collaboration of photographers, scientists, web developers and cartographers to create a record of ice on every continent before the glaciers disappeared. In twenty-first-century art the glacier is becoming a symbol of fragility and loss.

This sense of loss has been captured by glacier artists in different ways. Using water collected from 24 of Iceland's glaciers, in 2007 Roni Horn built tall glass columns filled with water to create a 'library of water' in a former library building in the Icelandic town of Stykkishólmur. For the Venice Art Biennial in 2007 sound artist Kalle Laar established a mobile telephone link to a glacier, through which the caller could hear the sound of the melting ice. Katie Paterson refroze glacier meltwater into the form of playable record discs, and played them on an acoustic turntable to create sounds that recalled the melting of the glaciers. Julia Calfee's project 'The Last Songs of the Glaciers', exhibited in 2010, develops this idea of capturing the sounds of the ice as it melts. Calfee has made more than a hundred recordings of the sounds or, as she calls them, the songs of melting glaciers that can be heard in the waters of the Rhine far, far downstream. The work of Canadian artist Linda Lang typifies this environmental engagement of glacier artists. Lang is a signature member of the Artists for Conservation group, and specializes in paintings that tell the story of how climate change impacts on polar regions. Lang also brings right up to date the long tradition of Canadian artists working in glaciated high latitudes. Lang's 2011 painting *Glacier Patterns on Bylot Island* is a reminder of the painting of Greenland by Lawren Harris that was mislabelled and appeared in error on a Canadian stamp. Perhaps Lang's painting, which is truly and genuinely of Canada's Bylot Island, deserves a stamp of its own.

Art as art, art as data, data as art

Some of the overlaps between art and science are immediate and direct. Projects such as the British Antarctic Survey's 'Data as Art' repurpose and re-present scientific data to create art that is simply visually impressive data.[6] By contrast, some artists consider their work to be quite explicitly an antidote or alternative to scientific work, rather than an adjunct. Artist Elizabeth Jackson describes her 'Glacier Project', which focuses on the careful extended study of the colours of glacial ice, as

> a visual study of glaciers as opposed to a scientific study – not as attempt to rework or reject a scientific approach, but simply to offer an alternative interpretation of glaciers. Something to provide poetry – poetry that science sometimes filters out in the pursuit of objective discourse.[7]

On the other hand some art becomes part of the science itself. Many present-day glacier artists describe their work as producing a record of glaciers before they retreat further or disappear entirely. Although the issue of their disappearance was not so much on their minds, previous generations of artists also made it their business to record the positions and characteristics of glaciers. The records that those artists produced, going back far beyond the invention of photography and in many locations before any detailed scientific monitoring, provide today's scientists with important data. Scientists such as Heinz Zumbühl, Wilfried Haeberli, Michael Zemp and others have described in detail how long histories of glacier advance and retreat, stretching back several hundred years, can be reconstructed from old paintings and drawings. Samuel Nussbaumer

So-called erratic blocks carried by glaciers have been a source of inspiration to many artists and a source of information for scientists. Rudolf Hentzi (1731–1798) (after Caspar Wolf, 1735–1783), *La Grosse Pierre sur le glacier de Vorderaar* (The Great Stone on the Glacier of Vorderaar), 1789–91, colour aquatint.

John Singer Sargent, *Schreckhorn, Eismeer* (from 'Splendid Mountain Watercolours' Sketchbook), 1870, watercolour and graphite.

and Heinz Zumbühl describe in a 2012 paper how drawings, paintings, prints, photographs and maps, as well as written accounts, were analysed to produce a history of the Bossons Glacier from the mid-eighteenth century up to the 1860s, and refer especially to the meticulous drawings by Jean-Antoine Linck, Samuel Birmann and Eugène Viollet-le-Duc that illustrated the glacier's vast advance and subsequent retreat during the nineteenth century.[8] Swiss landscape artist Samuel Birmann made many drawings and paintings of Alpine glaciers including the Lower Grindelwald Glacier, Switzerland, when it was in a relatively advanced position around 1820. Comparison of his pictures with earlier paintings, such as one made of the front of the same glacier in 1774 by Caspar Wolf, later photographs such as those by the Bisson brothers in the mid-1800s, and present-day observation, gives us an extremely detailed long-term record of the glacier's history.

Repeat photography is the most striking way of reporting the loss of glaciers over the last hundred years or so, and is today right on the interface between art and science where the expedition artists of the nineteenth century used to be. A number of high-profile exhibitions, films and publications have been based on long-term photographic records. The International Centre for Integrated Mountain Development (ICIMOD) mounted an exhibition, 'Changing Landscapes', combining photographs of the Everest region in the 1950s with repeat photographs collected from the same sites by mountain geographer Alton Byers in 2007. The exhibition was first unveiled in a small format at the Mount Everest Base Camp in April 2008, making it the highest photo exhibition in the world. At a similar time, David Breashears established the GlacierWorks organization, the declared mission of which was to document, educate and raise awareness about changes to Himalayan glaciers through art, science and exploration. GlacierWorks in association with the Asia Society mounted the exhibition 'Rivers of Ice – Vanishing Glaciers of the Greater Himalaya' based on repeat photography but expanding into a major monitoring and recording effort that can be followed not only through the exhibition but through film and the GlacierWorks website. 'Rivers of Ice' is one of a new breed of art-science interactions involving large, well-sponsored organizations at the interface between art and science, with a cultural, political and social element to their environmental interests. Another project in the same category and established around the same time is the 'Extreme Ice Survey' set up by James Balog. As a photojournalist and a trained earth scientist, Balog himself crossed that art-science boundary. The Extreme Ice Survey established time-lapse cameras at glaciers around the world and has spawned enormous and varied outputs including the film *Chasing Ice*, several books of glacier photographs and a website that packages them together with a strong environmental message. This is not art for art's sake. This is art with a message, art seeking sponsorship and art asking you to do something specific. One thing that has emerged through these projects is the particular value of time-lapse photography. Without repeat photography, without time-lapse photo-video, we would not see these changes in the same way. Art and science combine here to show us

overleaf:
An advertisement (*c.* 1889–91) for photographic prints of Alaska by Frank Jay Haynes (1853–1921).

Frank Jay Haynes, *Self-portrait at Glacier Bay, Alaska*, 1889–91, albumen silver print from glass negative.

✦ Alaska Views. ✦

BOUDOIR SIZE 5¼ X 8½ INCHES.

5001 Entrance to Wrangel Narrows.
5002 Frederick's Sound.
5003 Steamer Landing, Fort Wrangel.
5004 The Whale, "
5005 Totem Poles (The Bear.) "
5006 Totem Poles, "
5007 Totem Pole, "
5008 First Iceberg, Taku Inlet.
5009 Davidson Glacier, "
5010 Davidson Glacier, "
5011 Davidson Glacier, "
5012 Taku Glacier,
5013 " "
5014 " "
5015 " "
5016 " "
5017 Tredwell Mine, Douglas Island.
5018 From Deck of Steamer at Tredwell Mine.
5019 Juneau, Alaska.
5020 " " from deck of Steamer.
5021 Auk Glacier, from the Steamer.
5022 Pattison Glacier, " " "
5023 Chilkoot Range.
5024 Chilkaht Range.
5025 Chilkaht Peaks.
5026 Pyramid Bay, Chilkaht Inlet.
5027 Chilkaht, Alaska.
5028 Str. "Queen" in ice, Glacier Bay.
5029 " " approaching Muir Glacier.
5030 " " at Muir Glacier.
5031 Face of Muir Glacier, from Steamer.
5032 Face of Muir Glacier, from Steamer.
5033 " " " " from Steamer.
5034 " " " " from Steamer.
5035 Face of Muir Glacier, from Morain.
5036 " " " " from Morain.
5037 " " " " from Morain.
5038 " " " " from the top.
5039 Crevasse in Muir Glacier.
5040 " " " "
5041 Top of the Muir Glacier.
5042 Ice Peaks on Muir Glacier.
5043 Tourists on Muir Glacier.
5044 Glacier Bay, from Muir Glacier.
5045 Muir Glacier and Bay.
5046 " " " "
5047 " " " "
5048 " " " "
5049 Glacier Bay.
5050 " "
5051 " "
5052 " "
5053 " "
5054 " "
5055 Sitka, Alaska, from the Steamer.
5056 Sitka Harbor, " " "
5057 Indian Avenue, Sitka.
5058 Baranoff Castle, "
5059 Greek Church and Trading Store, Sitka.
5060 Ocean View from Sitka.

PHOTOGRAPHED AND PUBLISHED BY

F. Jay Haynes & Bro., Official Photographers N.P.R.R.

392 Jackson St., Cor. 6th,

St. Paul, Minn.

YELLOWSTONE PARK AND NORTHERN PACIFIC VIEWS.

something new and valuable. The purpose of both art and science is to help us to see more and to see more clearly.

Modern art-science collaborations

Some art-science collaborations have been deliberately conscious of the distance separating these two approaches. Siân Ede's book *Art and Science* asks: 'Is science the new art? Scientists weave incredible stories, invent wild hypotheses and ask difficult questions about the meaning of life . . . Contemporary scientists frequently talk about "beauty" and "elegance"; artists hardly ever do.'[9] Art and science are not the same thing, but the boundary areas between them have proved rich ground for smaller projects independent of the corporate or political sponsorship typical of big environmental glacier megaprojects. There is a grass-roots development of glacier art steering a careful path along the border of science without trying to become science, at the border of politics without becoming politics.

Anna McKee is a Seattle-based artist interested in what she calls 'memories that accumulate in the physical world; specifically where human history intersects a longer time span'. As part of that interest in time and memory she has talked to scientists working on glaciers and the Earth's history, visited their field camps and spent time drawing ice cores that have been collected from Antarctica. The outcome, for McKee, is not some uncomfortable hybrid art-science political-environmental message, but a personal artistic response to the things she has seen. The scientists were measuring chemical composition of the ice, calculating ages and ice-flow trajectories, finding evidence of past atmospheres in gas bubbles. McKee writes: 'I imagined all of the secrets drifting up into dozens of laboratories as scientists release these bubbles during their experiments . . . Scientists are looking for isotopes, trace elements and biological matter. I wondered about penguin breath, remnants from long ago polar expeditions and gases from Mt Vesuvius.'[10] This is not art as science or science as art, but art and science as two different ways of seeing; two different approaches to knowing. Such

For an exhibition in 2010 about glaciers and the ice age, artist Miriam Burke created permanent snowballs by assembling sheets of glass on which falling snowflakes had been caught in superglue.

collaborations, or juxtapositions, have resulted in a number of exhibitions in which the artist's and the scientist's responses to the same inspiration have been placed side by side. Miriam Burke and Peter Knight's exhibition 'Know This Place for the First Time' placed Knight's glacier science paraphernalia alongside Burke's artwork. Artist Emma Stibbon visited scientist Giles Brown's old glaciological field sites in the European Alps and together they put together an exhibition, 'Glacial Shift', which placed together as contrasting records Stibbon's art and Brown's scientific data from the same sites. Stibbon noticed the glaciers' effect not only on the landscape but on the imagination. The value of collaborations and exhibitions of this kind is not that the artist and the scientist arrive in the same place, but that from the same starting point they go by different routes and arrive in different places.

These examples point towards an important area of art-science collaboration that aims not to merge the two, but to go beyond both, each inspired by the other to look at new things or to see old things in new ways. To the scientist, Earth is geology; but to the artist, Earth is art. There are many exhibitions, websites and books presenting abstract images of Earth's surface from space or from very close up so that things seem out of context. Earlier in this book we looked at how glacial erratics, those rocks carried by ice away from their home locations, had inspired early geologists in working out the history of ice ages. Even those glacial erratics can be seen differently through the eyes of an artist. Around 2012, the American photographer Fritz Hoffmann conducted a countrywide survey of erratic boulders, photographing them in context with the modern world that has grown up around them. He was interested especially in the idea of time. For Hoffmann, what was remarkable about these rocks was the way they stayed put, motionless, where they had been for thousands of years, while people were rushing busily past to go about their daily lives: his photographs captured the rocks and the people existing at different rates of motion.

Earth is made into art – glacier is made into art – by the artist's choice of what to notice.

8 Glacier Stories and Songs . . . Once upon a Glacier

As symbols in the metaphorical language of art, glaciers have many meanings. They stand for the grand, the romantic, the slow but inexorable, the barren, the stoic, the pure, the fragile. They are the ultimate wilderness, the great adventure, our planet's history. Superman's Fortress of Solitude. Frankenstein's monster confronting his maker. The glacier is both hero and victim in the twenty-first-century environmental narrative, and arch villain in a string of climate-disaster movies. There are glaciers in W. H. Auden's poetry, in Jules Verne's fiction and in the Harry Potter Quidditch World Cup video game, produced by EA (Electronic Arts Inc.). In 2000 film-maker Ruth Meyer made the dance film *Breath Crystal* in memory of her grandparents who died in Auschwitz. It was a choreographic interpretation of Paul Celan's holocaust commemoration poem 'Weggebeist' delivering a message about human fragility and perseverance through a dancer's journey across the surface of a glacier. People have used glaciers to say a lot of different things in a lot of different ways.

The musician Ned Selfe grew up in the deep south of the USA before moving to Hawaii. His main instrument is the steel guitar. In 1995, when he produced his first album, he chose the title *Glaciers Come, Glaciers Go*. It didn't seem an obvious choice of title for a new age rock-jazz album out of Hawaii, so I wrote to ask him why. He explained that he wanted an image that would allude to the transitory nature of human consciousness: we all tend to inflate our current problem or obsession into a giant megalith that seems forever unchanging and all-consuming, he said, when in fact it will soon melt and fade into the next thing that will occupy our thoughts. Selfe told me that the idea of using the glacier for that metaphor came from his reading of M. Scott Peck's book *The Road Less Travelled*. Writing about how we choose a map for our life, Peck wrote:

> The biggest problem of map-making is not that we have to start from scratch, but that if our maps are to be accurate we have to continually revise them. The world itself is constantly changing. Glaciers come, glaciers go. Cultures come, cultures go . . . the vantage point from which we view the world is constantly and quite rapidly changing . . . we must continually revise our maps.[1]

Glaciers feature in the traditional literature and oral histories of many cultures. In the Norse creation myth *Völuspá*, the universe began in the clash between fire and ice. Ice, the evil, burst out from a poisoned well. Fire, the good, met the ice and formed a giant who was killed by the gods. The Earth was created from the giant's carcass. In another version of the myth, Ginnungagap, the mighty void that existed before creation, was filled from the north by the cold of the frozen primordial realm of Niflheim and from the south by fire from the realm of Muspelheim. From the meeting of ice and fire came the steam out of which everything was created.[2]

In the European Alps many traditional glacier stories are preoccupied with catastrophes and disasters, reflecting the history of glaciers as a threat to human activity.[3] In one story from Switzerland, the Aletsch Glacier, the longest glacier in the Alps, was formed by the freezing of the last waters of the biblical flood. The freezing of the glacier imprisoned evil within the ice so that the people could live in peace. Whenever the Aletsch retreats and its waters are released, the evil spirits are freed in the 'ghost hour', spreading terror to the region. When the glacier readvances the evil is locked up again.

The oldest glacier-related stories have been passed down by oral tradition in Australia. Studying traditional stories that encode memories of extreme natural events such as floods, geographer Patrick Nunn has identified stories passed down through hundreds of generations of Aboriginal Australians as a hi-fidelity oral history of the landscape at the end of the ice age when the sea level was rising.[4] Some stories include details of locations that vanished under water 10,000 years ago. A story told by the Tiwi people of northern Australia describes the creation of Bathurst and Melville islands. An old woman is said to have crawled across dry land between the islands, followed by a flow of water. In reality, that separation of the islands from the mainland took place about 9,000 years ago. In the southeast of Australia, stories refer to an area on the coast of Victoria as rich ground for hunting kangaroo. The area is now Port Phillip Bay, flooded by postglacial sea-level change 8,000 years ago. Flood stories are common in the traditions of cultures around the world, including Noah's flood in the Bible. These stories may derive from the same global events that the Australian Aboriginal stories recount in more literal detail.

In the hazard-strewn legends of the European Alps, people seem to be distinctly separate from, and at odds with, the landscape. Glaciers are the threat, the other. But in some mythologies the land and the people are connected parts of a whole. Rebecca Solnit, in her *Field Guide to Getting Lost*, describes the Wintu people of north-central California, who don't use the terms 'left' or 'right' to describe which hand they point with or which foot is on the bottom step, but use the directions north, south and so on.[5] I may point at the glacier with my west arm. If I turn to face the other direction, it becomes my east arm. My description of myself depends not just on me but also on my position in the landscape. The Wintu set themselves within their surroundings rather than against them. Such differences in attitude are reflected in traditional stories. The cultural historian Julie Cruikshank has described the place of glaciers in oral traditions of North American indigenous peoples, for whom the natural landscape is not inanimate.[6] Languages such as Tlingit and Athapascan use words for the behaviour of glaciers that allow for deliberate action, and for responses to the actions of others: glaciers can make moral judgements and deliver punishments. For example, the Tlingit people of the Pacific Northwest coast have a story about a girl named Kaasteen who insulted the glacier, and the glacier came down and destroyed her village and drove the people out of the bay where they lived.

Glaciers featured in traditional histories because they featured in people's lives. The same is true of some more recent writing. Novels and films set in glaciated areas feature glaciers as inevitable parts of the background, and in some the glaciers become actors in the story, drawing upon the iconography of glaciers to add something extra to the plot. Halldór Laxness won the Nobel Prize for Literature 1955 for novels about the people of Iceland. In Laxness's stories, as in those that Cruikshank reported from the Pacific Northwest, glaciers show intention and personality. In *Under the Glacier* Laxness draws on the idea of glaciers as calm, unhurried, stoic and achieving things by steady, slow progress. When the bishop's emissary asks the pastor about giving sermons, the pastor replies: 'Oh no, better to be silent. That

J.M.W. Turner, *Mer de Glace, Valley of Chamouni-Savoy* (*Liber studiorum*, part x, plate 50), 23 May 1812, etching and mezzotint. Turner's picture reflects the 19th-century Romantic idea of the Alps that inspired Mary Shelley when she created her story *Frankenstein*.

is what the glacier does. That is what the lilies of the field do.' This is glacier as role model.[7]

In Thomas Wharton's novel *Icefields* (1995) a character slips into a crevasse on a glacier and before he is rescued he sees embedded in the ice the figure of an angel, its wings outstretched. The idea that glaciers were protagonists in creation myths and are now ancient conveyors of items deposited long ago lends plausibility to both a magical realist interpretation of Wharton's story and to the idea that a character living in the remote glacial wilderness might suffer from psychological delusions. Who can say, the story seems to suggest, what we might see when we gaze into the ancient depths?

Glaciers emerged from the European Romantic movement in the eighteenth century as metaphors for the awe-inspiring. Along with mountains, oceans and storms, glaciers epitomize the notion of the sublime: that combination of terror and beauty that lay at the heart of Romanticism. Until the late eighteenth century, wild mountainous areas such as the Alps were seen primarily as a nuisance to travellers, an unpleasant disfiguration of landscape, and a hiding place for robbers and evil spirits. But as the Romantic movement grew, the Alps became a destination in their own right. In 1816 the English Romantic authors Percy Bysshe Shelley and Mary Shelley spent the summer in the area around Chamonix, staying with the poet Lord Byron. During this visit Mary Shelley started writing her novel *Frankenstein*. Her description of a glacier in that story is typical of the Romantic view of terrible but beautiful nature:

> The abrupt sides of vast mountains were before me; the icy wall of the glacier overhung me; a few shattered pines were scattered around; and the solemn silence of this glorious presence-chamber of imperial nature was broken only by the brawling waves or the fall of some vast fragment, the thunder sound of the avalanche or the cracking, reverberated along the mountains, of the accumulated ice, which, through the silent working of immutable laws, was ever and anon rent and torn, as if it had been a plaything in their hands.[8]

Mer de Glace.

On that same visit, Percy Bysshe Shelley described glaciers in his poem 'Mont Blanc'. He said the poem

> was composed under the immediate impression of the deep and powerful feelings excited by the objects which it attempts to describe; and, as an undisciplined overflowing of the soul, rests its claim to approbation on an attempt to imitate the untamable wildness and inaccessible solemnity from which those feelings sprang.[9]

Byron, Shelley and other poets such as Wordsworth and Coleridge were pre-eminent among writers calling out a new message about these wild landscapes. Wilderness was no longer forsaken ground, but sacred ground where the traveller could see either God, or their own souls, reflected in nature. In *A Tramp Abroad* (1880), Mark Twain called the Alps 'the visible throne of God'. Byron himself wrote:

> Above me are the Alps,
> The palaces of Nature, whose vast walls,
> Have pinnacled in clouds their snowy scalps,
> And throned Eternity in icy halls
> Of cold sublimity, where forms and falls
> The avalanche – the thunderbolt of snow!
> All that expands the spirit, yet appals,
> Gather around these summits, as to show
> How Earth may pierce to Heaven, yet leave vain
> man below.[10]

In successive waves the Grand Tourists, the poets and artists, and then the climbers and mountaineers, were each enthralled and had the Alps – as they were shown to them by the Romantics – stamped onto their individual and national psyches. Other writers found similar inspiration in other mountains and other glaciers. For Pushkin and Lermontov, for example, the Caucasus Mountains offered material for both artistic and political Romanticism as Imperial Russia conquered the mountains and their native inhabitants. By the start of the twentieth century, glaciers were firmly established in the lexicon of literary shorthand.

W. H. Auden's poem 'As I Walked Out One Evening' includes the lines 'The glacier knocks in the cupboard,/ The desert sighs in the bed,/ And the crack in the tea-cup opens/ A lane to the land of the dead', and it is easy for readers to recognize the glacier as a symbol for the ancient and barren. The glacier and the desert clearly indicate the slow march of time towards desolation. Auden's glacier in the cupboard also draws on the obviously non-domestic scale of a glacier: the glacier here implies vast territories of vacant space within this cupboard, or a link to an imaginary or metaphorical territory, like a dark version of C. S. Lewis's wardrobe opening onto Narnia. The reference to the glacier calls up impressions of both size and antiquity. Like the trees that poet Odysseus Elytis describes as, 'the musical notation of another world . . . the very near and yet unseen',[11] Auden's glacier reminds readers who understand the shorthand that we see much less in our everyday view than is really there. Seamus Heaney, too, uses glaciers as metaphors. W. B. Yeats referred to 'emblems of adversity' representing a malevolent natural or supernatural environment, and many poets have used glaciers as metaphors in that way. In his poem 'Funeral Rites' Seamus Heaney described a funeral as a black glacier, invoking the glacier to denote the slow, the desolate, the implacable. Describing the Arctic December, Alaskan poet Tom Sexton described the long winter night as moving slowly, like a black glacier.

There is a long tradition of poets being involved in scientific expeditions. The Harriman expedition to Alaska in 1899, which included botanists, entomologists and geologists, also included author John Burroughs and poet Charles Augustus Keeler. Keeler's 1890 poem 'To an Alaskan Glacier' includes the strikingly scientific metaphor of the sea as mother to a glacier, and of the ice moving from the mountains back to the sea where the water returns to its source:

> Out of the cloud-world sweeps thy awful form,
> Vast frozen river, fostered by the storm

Upon the drear peak's snow-encumbered crest,
Thy sides deep grinding in the mountain's breast
As down its slopes thou plowest to the sea
To leap into thy mother's arms, and be
There cradled into nothingness.

Keeler's poem includes some sound basics of glacier science: formation of the glacier from snow crystals; transfer of ice back to the ocean; and erosion of the landscape by the moving ice. It also follows the Romantic approach of granting the glacier feelings to match the sublime landscape, 'roaring like a God in fierce dismay . . . eager in one mighty throe to leap into the sea and end thy woe'. But in one critical point the scientific and the poetic overlap, as Keeler refers to the age of the slowly moving ice and the length of time over which it makes its mark on the landscape:

How slow,
How imperceptible, thy ceaseless flow,
As one with an eternity unspent
Wherein to round thy task of wonderment!

and

Years, centuries and eons thou hast known,
Waxing and waning in the wilds alone.[12]

The notion of what we now call 'deep time' emerged from the work of geologists and biologists in the eighteenth and nineteenth centuries, overcoming earlier beliefs in the very limited age of the Earth deduced from biblical evidence. The work of scientists such as Darwin with his theory of evolution, James Hutton in his *Theory of the Earth* and Charles Lyell with *Principles of Geology* was echoed in the work of poets such as Tennyson with his 'In Memoriam'. The struggle to reconcile our day to day experience, our mortality and our faith with the emerging recognition of Earth's great age and the scale of the universe were challenges for poets and scientists alike.

The tradition of combining poetry, art and science is continued today by the Cape Farewell expeditions, which have included poets in their roster of artists. Cape Farewell also teamed up with the Poetry Society to develop a Youth Poetics programme in which hundreds of young poets submitted poems about climate change. Poets have also been installed 'in residence' in academic institutions to work alongside glacier scientists. Allyson Hallett was funded by the Leverhulme Trust to be poet in residence in geography at the University of Exeter in 2010. She worked with geographer Chris Caseldine on a project based around a student field trip to Iceland. The output combined Hallett's poems, student photographs of the landscape and scientific essays by Caseldine.

Some glacier writing has stemmed from the experiences of authors working professionally in glaciated regions. The poet Maurice Chappaz took a job at the construction site of the Grande Dixence Dam in Switzerland. The dam was constructed between 1950 and 1964 as part of a system to collect glacier meltwater for use in hydroelectric power generation. Chappaz is known also as the author of *La Haute Route*, a guide to the famous skiing route from Mont Blanc to Zermatt, which has been referred to as the literary bible of ski mountaineering. Working on the construction of the dam put Chappaz at the heart of the conflict between preserving the natural beauty of Valais and exploiting its resources, and gave him a new perspective on the value of 'progress'. Based on his experience he wrote critically about the effects the tourist industry had on Valais. The story of Chappaz and his engagement with the environmental movement was made into a documentary film, *Alpenglühen – Der weiße Weg: Maurice Chappaz und die Haute Route*, by Sigrid Esslinger.

German film-makers have a long association with the Alps. In the 1920s German mountain film, or *Bergfilme*, became established as a genre associated, like the American western, with a particular location and scenery. Also like the western, *Bergfilme* often involved the landscape itself as an important part of the action, and a hero being transformed by some experience that

Perito Moreno Glacier, Argentina.

took place within the landscape. One of the most prominent film-makers in the German mountain film genre was Arnold Fanck, who was not only a film pioneer but held a PhD in geology. Leni Riefenstahl's film-making career started when she convinced Fanck to give her a role in his film *The Holy Mountain* (1926). One of their most successful collaborations was *The White Hell of Pitz Palu* (1929) (*Die weisse Hölle vom Piz Palü*), the story of a man searching the mountain Piz Palü for the body of his wife, who was swept onto the glacier in an avalanche on their honeymoon. One of the avalanches in the film was a real avalanche that threatened the safety of the cast and crew, and Leni

Riefenstahl was affected by frostbite. German mountain film waned in popularity and disappeared in the 1930s, and today the term 'mountain film' is more often associated with documentary films based on climbing expeditions, and with the many 'mountain film festivals' that are organized all over the world.

In more recent cinema, glaciers feature in a relatively small share of action adventure and disaster movies. Volcanoes, earthquakes and killer versions of everyday creatures top the list for natural-disaster cinema high spots, but glacier cinema has at least attempted to ride the wave of environmentally aware Hollywood output. Many of the presentations of glaciers in Hollywood cinema, like most Hollywood versions of the natural world, are patently ridiculous. As a glaciologist with an interest in environmental change it is hard to know whether to laugh or cry at *The Day after Tomorrow* in which a new ice age starts more or less overnight, or films like *2012: Ice Age* which uses the tagline 'A volcanic eruption in Iceland sends a glacier towards North America, causing everything in its path to freeze.' In the official

The glaciated mountain Piz Palü in the Bernina Range of the Alps between Switzerland and Italy, which was the setting for the 1929 film *Die weisse Hölle vom Piz Palü* (The White Hell of Pitz Palu).

trailer a character looks at the darkening sky and says, 'It's getting cold, fast.' 'What can do that?' she asks. 'Glaciers,' says her companion, 'really, really fast glaciers.' The science is nonsense. In fact the scientific underpinning to the Twentieth Century Fox *Ice Age* series of animated children's films is much stronger. The cartoon creatures migrate south to escape the advancing cold at the start of an ice age. There is sound palaeozoology in that premise. In *Ice Age 2: The Meltdown* melting ice sheets at the end of an ice age cause catastrophic flooding. At the start of the film, after the sabre-toothed squirrel has seen water spurting from a glacier, our heroes discover that melting ice has formed a great lake that is held back from flooding their valley only by a glacier dam. The whole film is based on the ice-dammed lakes and glacier outburst floods that really did occur at the end of the last ice age. There are academic textbooks on exactly the same topic. The *Ice Age* animated movies make a much more educationally valuable contribution to the public understanding of science than films like *The Day after Tomorrow*.

A tourist boat provides a close-up view of the glacier lagoon in front of Breiðamerkurjökull, one of Iceland's premier filming locations for TV and cinema.

In cinema, as in other media, landscape is commonly used as a metaphor for character, for underlying themes in the storyline or for the human condition. Glaciers usually symbolize the cold, the austere, the inhospitable, but sometimes also the pure or pristine. In both respects, the glacial landscape is the ideal home, refuge or training ground for a superhero. The scenes in *Batman Begins* (2005), where Batman trains to develop his skills, take place beside glaciers (supposedly in Bhutan, but filmed in Iceland). Superman's Fortress of Solitude – although it takes different forms in different versions of the story in print and on screen – is typically icy and Arctic in design. James Bond, Lara Croft . . . a long roster of characters has been filmed against the backdrop of Iceland's glaciers. When the screenplay for the TV series *Game of Thrones* called for a murderously cold no man's land to the north, the crew headed for Iceland. The planet Hoth in *Star Wars: The Empire Strikes Back* was Hardangerjøkulen Glacier in Norway.

The classic Hollywood western is one genre in which the landscape typically plays an integral role in the story. The whole premise of the genre is in the progress of frontier: land disputes, establishing transport links, setting up settlements in the wilderness. It is an essentially geographical genre. Particular directors are closely associated with specific landscapes where they choose to film. Director Anthony Mann made a series of westerns with James Stewart set in river and mountain country stretching up through forests to the snowline. In *The Far Country* the action is set in the Yukon in gold-rush days and unusually, but wonderfully, the film combines traditional western action with glacier landscapes. The film was shot on location around Jasper National Park, and the opening titles roll over footage of the Athabasca Glacier. The landscape is primarily a background and setting, but in this case a very active setting, with glacier hazards coming to the fore as characters crossing in front of a glacier are struck by an ice fall.

In the 1948 film *Scott of the Antarctic*, stock footage of Graham Land in the Antarctic Peninsula was intercut with film from locations in Norway and Switzerland to represent the film's bleak

landscape. Ralph Vaughan Williams's musical score for the film was composed to match the desolate visual landscape. Vaughan Williams reworked the score in 1952 as *Sinfonia Antartica*, his seventh symphony. In music, the same issues apply to the use of glaciers and glacial landscapes as icons or motifs as apply in other media. Music can describe the physical and metaphorical properties of glaciers, take advantage of that same metaphorical

The surface of the glacier Svínafellsjökull in southern Iceland was used as a filming location to represent the ice planet in the 2014 movie *Interstellar*.

language used by writers to reach an audience, and engage with the environmental conversation of which glaciers are a central part. In 2016 the composer and pianist Ludovico Einaudi, working with Greenpeace, released a video of himself performing a specially written piece of music called 'Elegy for the Arctic' on a grand piano afloat in front of a calving glacier. The release was timed to coincide with a political debate about whether to allow further resource exploitation of the Arctic. 'Raise your voice to save the Arctic,' said the caption for the video, as ice from the melting glacier collapsed as a cliché, but a powerful one, into the iceberg-strewn sea.

Peter Broderick's debut album *Float* (2008) included a track called 'A Glacier'. I asked Broderick why he had chosen that name, and he explained that it was all about 'the apparent simplicity of them but also this sort of grand mysteriousness . . . with the music I just really loved this image of an enormous block of ice slowly creeping along. I had that picture in my head while writing most the music for that album.'[13] For a song called 'Another Glacier' on the same album, Broderick told me that he wrote the lyrics from the glacier's point of view: 'The glacier is uncovering, slowly, new pieces of earth . . . seeing new pieces of ground emerging and saying to itself: "Am I moving? Or just melting?"'[14] The rock band So Many Dynamos have a more environmental edge in their choice of 'Glaciers' for one of their titles. Despite the melting glaciers, their lyrics proclaim, people will go on living, as they always have done. Similarly, the band British Sea Power's 2005 release *Open Season* includes the track 'Oh Larsen B' about the collapse of the Antarctic Larsen B ice shelf. The lyrics describe the melting ice desalinating the sea, and the ice shelf coming to the end of its 12,000-year life. The song sounds almost as though it could have been an experimental collaboration between scientists and musicians. Such collaborations have taken place. The organizers of the July 2012 Scientific Committee on Antarctic Research Open Science Conference in Oregon invited participants to join in with a series of musical events held in conjunction with the conference. The aim was for Antarctic researchers to use their artistic talents to communicate scientific ideas to the public in an accessible form but without compromising accuracy or validity. The difference between this and other art-science crossovers is that this call was not for artists to collaborate with or be inspired by scientists, but for scientists to produce their own artistic work based on their own science.

There are a lot of different ways of telling stories about glaciers, whether in literature, film or music. Historian W. G. Hoskins argued that poets make the best topographers.[15] I would argue that there are many different types of poet.

9 Adventure, Exploration, Inspiration

Where the glacier meets the sky, the land ceases to be earthly, and the earth becomes one with the heavens.
Halldór Laxness[1]

In the past, people generally only interacted with glaciers if they had to. Nowadays, a lot of people want to interact with glaciers just because they can. Some of these people are the inheritors of a tourist tradition stretching back to the European eighteenth-century Grand Tour and the birth of the Romantic movement, when the Alps were transformed from being a terrible obstacle into a destination in their own right. Other modern-day glacier enthusiasts inherit a cultural memory of the great days of polar and mountain exploration, following in the footsteps of their Victorian or Edwardian heroes and pitting themselves against the adversity of remote wilderness. For some, a visit to a glacier is an environmental pilgrimage: a 'last chance to see' tick-box on a green agenda. Others perhaps just have a gap year, a round-the-world ticket and a camera. Many people who visit glaciers are not interested in ice ages or the history of science; they are not even too concerned about disappearing ice. They mainly want to have an 'experience'; to have fun. Some people go to Disneyworld, some people sit on the beach with a beer, some people go to a glacier. Some of them go with climbing boots, bungee ropes, skis or kayaks. Some of them go with a ring, a best man and a few bridesmaids. Glaciers offer one of the world's great adventures, but in the twenty-first century people can take their glacier adventures in many different flavours.

In the past people encountered adventure often in the name of duty, science or exploration. The most famous of the great expeditions of the past, which are remembered now most often for the spectacular adventure they involved, were established as part of some loftier project for science or for nation. Fridtjof Nansen's first crossing of Greenland in 1888 – an adventure now repeated by amateur adventurers every year – was primarily an expedition of science and of geographical exploration. When Nansen set out on his journey, it was not even known whether the whole of the interior of Greenland was ice-covered. Some people suggested that there would be open land in the centre, suitable for colonization. There were no satellite images, no aircraft reconnaissance. Nobody had ever done it before. Nansen's plans were subject to scientific referees and the outcome of his 'adventure' would be of historical significance. Many of our modern adventure holidays and excursions are imitations of, or homages to, these historic predecessors. The people who ski across Greenland today for fun do not, on the whole, make any

Tourists visiting a Norwegian glacier in the early 20th century.

great contributions to science or to our knowledge of the world. There are several adventure-tourism companies that run guided ski crossings of the ice sheet. The typical fee is around £8,000 per person. When Nansen embarked on the first Greenland crossing, setting off from the east coast into the unknown interior, he declared, 'Death or the west coast!' I don't think any of today's commercial tour guides use that as their motto for clients. When Scott and his team sacrificed their lives in their attempt to be first to the South Pole in Antarctica, they did so not in the name of adventure (although adventure it was) they did so in the name of exploration and science. Nobody had been there. We did not know about the geology, the biology, the meteorology of Antarctica: their expedition was to discover something new and it was done in its particular way not because using horses and dogs, and man-hauling sledges, was 'fun' or an 'experience' but because at that time there was no other way to do it. Today, one can fly to the South Pole, live in a climate-controlled building there, and travel around Antarctica with helicopters and motorized skidoos. We do not need to man-haul sledges across the great barrier, up the Beardmore Glacier and across the polar plateau. And yet people still do it anyway. Glaciers provide a hostile environment where we can set ourselves extraordinary challenges.

Sometimes these challenges are unashamedly in the name of adventure, but often they are careful to include a scientific or environmental justification, and the justification usually features prominently in the literature seeking sponsorship from potential benefactors. Some modern adventures are explicitly repeating historical journeys. In 2001 a group of artists, scientists and writers retraced the route of the 1899 Harriman Alaska expedition. That original expedition, set up by railroad tycoon Edward Harriman, included John Muir, founder of the Sierra Club, who was an acknowledged expert on Alaskan glaciers and at the time of the expedition already had a glacier named after him. Muir took the opportunity during the expedition to name another glacier after the expedition's sponsor and leader. In Antarctica, the Beardmore Glacier was given its name by Ernest Shackleton in honour of one of the sponsors of his 1908 Nimrod expedition.

Scientific endeavour has not been the only motivation or excuse for adventure in glacier country. Ever since the Romantics opened up the European Alps as a potential tourist destination, mountains and their glaciers have provided a way for visitors to experience something of the nature of heroic exploration even without the call of military or scientific service. One of the companies offering guided ski trips across Greenland describes their trip as 'polar travel in the romantic and demanding way of early explorers . . . This will be a spiritual journey as much as a physical journey . . . a long expedition requiring patience, stamina, mental fortitude and strong will power.'[2] For many people, the point of a glacier expedition is to challenge themselves, to do something very difficult and potentially dangerous. For the original Romantics, there had to be an element of danger for there to be an experience of the sublime. For the modern-day adventurer, a glacier adds an extra layer of danger to any land-based adventure. Climbing a mountain is one thing, but climbing a glacier is something more. Crossing a plateau may be an expedition, but crossing a plateau icefield is something else.

Jim Ring's book *How the English Made the Alps* describes part of the history of the growth of this kind of adventure tourism.[3] The mid-nineteenth century saw an explosion in mountain tourism. With the foundation of the Alpine Club, and the development of skiing as a popular holiday activity as well as a sport, the transformation of the Alps led to the progressive erosion of the very elements that had initially attracted the Romantics. When Byron saw the sublime and inspiring Alpine landscape, there were no tourist busses, no climbing expeditions, no scenic rail tours. We can still go and experience the physical, picturesque majesty of the Alps, but things have changed. For the early Romantics at the start of the nineteenth century the Alps were a sublime and remote wilderness. In 2014 the Zermatt-Matterhorn Tourism Report recorded more than 2 million overnight hotel bookings for that area alone.[4] At Fox Glacier in New Zealand, the local glacier guide company claims to have escorted more than 1 million visitors onto the ice since 1974. In Iceland, Alaska, New Zealand and elsewhere businesses have

A climber approaching the summit of the Matterhorn, Switzerland.

been set up based entirely on a booming market for weddings held on glaciers, including companies that will ferry in the wedding party and guests by helicopter to some remote and spectacular location. In 2016 the first wedding was held actually inside the Langjökull Glacier in Iceland, in a 500-m-long (1,640-ft) tunnel drilled under the ice. If Las Vegas is not to your taste, there are plenty of glacier alternatives for your happy day. In the 2014–15 tourist season, nearly 40,000 people travelled to Antarctica. Some merely cruised past, admiring the scenery and the wildlife, but in that season more than 27,000 tourists landed in Antarctica. Twenty-seven thousand.

For modern adventure tourists in popular resort areas, glacier environments have become one of the world's great adventure theme parks. Being in touch with the Earth, at one with the elements, is the basis for many cult sports. Glaciers offer many opportunities of that type. They can be booked online with a few clicks, and accessed with an international flight from anywhere in the world. Glacier adventure need no longer be remote for any traveller. From my sitting room right now I could book a guided walk into 'a magical frozen world' on an Icelandic glacier for about U.S.$100, or an afternoon of glacier sea-kayaking on the Tasman Glacier Lake in New Zealand for NZ$145. Or for U.S.$400 I could have a whole day kayaking up an Alaskan fjord to see the Aialik Glacier. 'Listen to the sound of avalanches thundering off of the nearby Mt Sefton as you look out for small icebergs and possibly glacier calving,' cries the advertisement for a kayaking afternoon at Mueller Glacier in New Zealand. 'Gaze in awe at the nearby slopes of Aoraki-Mt Cook as you explore this constantly changing and evolving lake.' If I don't fancy the water, I can book a day's heli-skiing in the same area of New Zealand for about NZ$1,000. For just NZ$500 I could enjoy an introduction to ice climbing on Fox Glacier, including two helicopter flights and professional instruction. Or if I prefer a less adventurous adventure, there are plenty of glaciers where companies have established bus routes right onto the ice. For less than U.S.$60 I could book my seat on an eighty-minute bus trip onto the Athabasca Glacier in British Columbia, and the advertisement tells me that I may safely step out onto the glacier to take photographs if I wish. 'The Great Wall of China, the Statue of Liberty, the Pyramids of Egypt; every corner of the world has its must-see attraction . . . in the Canadian Rockies, it's the Glacier Adventure.' In Iceland I can take a ride in an 'eight-wheel drive glacial super truck' onto the Langjökull Glacier, and at Jökulsárlón Lagoon on Iceland's south coast I can ride a bus that turns into a boat as it drives out into the water at the front of one of Iceland's most spectacular glaciers. This is the glacier where scenes have been filmed for James Bond, Batman and other adventure-filled stories, and such mythical adventures have fuelled a whole new breed of tourism as fans retrace the steps of the stars and visit the locations of their favourite movie scenes. 'Head out to the magnificent glacial lagoon Jokulsarlon,' says the advertising website, 'and follow in the footsteps of James Bond in "Die Another Day" and Lara Croft in "Tomb Raider".'

All of these organized adventures are careful in their advertising to combine the hint of some danger with a categorical assurance of safety. Just as for the Romantic sublime, modern-day tourists want thrills but they don't want genuine danger. Rather like a roller-coaster theme park, the ideal glacier theme park is just a little bit scary but completely safe. In 2015 'the world's first party underneath a glacier', with DJ performances and cocktails, was advertised as part of Iceland's 'secret solstice festival', but safety first was a priority, with the ticketing information stipulating that 'In the interest of keeping all guests of ICERAVE safe while under the glacier, no outside alcohol will be allowed during the ICERAVE tour or performance, and all guests are limited to a maximum of two drinks (provided as part of the package).' There is more than one reason why a party underneath a glacier may not be everybody's idea of a party.

Increasingly, glacier tourism is becoming associated with environmental issues. For many people, a trip to see glaciers is

A cruise ship taking tourists to view the rapidly retreating glaciers in Glacier Bay, Alaska.

Railway posters of the late 19th and early 20th centuries frequently drew on glaciated landscapes of the European Alps or the North American Rockies to advertise their destinations. This poster by Abel Faivre dates from 1905.

like a trip to see endangered creatures in the zoo. This has been encouraged by the widespread use of retreating glaciers as the poster children of environmental change, and by the recognition that in some of the best-known areas associated with glacier tourism the glaciers are indeed on the brink of extinction. Historical maps of Glacier National Park in Montana indicate that there were about 150 glaciers there in the mid-1800s. By the early twenty-first century there were only 25 active glaciers left, and current predictions are that those will disappear well before the middle of the century. Glaciers really are disappearing, and they are disappearing most quickly from areas such as the mid-latitude mountains, where they have until now been most accessible. The glaciers with which we are most familiar and which we can most easily visit are those that are about to vanish. See them while you can, suggests the advertising. Organizations often take advantage of this glacier-environment connection in their marketing, making the assumption that the kind of visitor who wants to see a glacier may well be the kind of visitor who wants to save the planet. One company in Alaska promises that for every person who books one of their glacier kayak trips they will donate $2 to help fund wind turbine installations and renewable energy education at schools across Alaska. Two dollars.

Even when the glaciers are gone their landscapes provide some opportunity for tourism. The English Lake District, so important to that early Romantic movement which initiated the development of wilderness tourism, has had no glaciers during human history but retains the imprint of their presence. The U-shaped valleys, sharp ridges and boulder fields left behind

Glacier d'Argentiere, Chamonix Valley, France – prudent advice to tourists.

Glacier tourism in the Chamonix Valley, France.

by glaciers can be impressive and inspirational in their own right. Even close to surviving glaciers, their ancient impact on landscape is celebrated and exploited for tourists. In Lucerne, Switzerland, the so-called Glacier Garden is a park which includes glacial meltwater potholes, striated boulders and ice-polished bedrock. There is a glacier museum, educational tours and a garden of plants from the ice age. If the glacial rocks and plants are not sufficiently tempting, there is a mirror maze, an events venue, and the oldest mountain relief model in the world.

One refreshing counterpoint to glacier tourism is the approach taken in Wisconsin, where an area of land that seems to have escaped glaciation in the last ice age is celebrated and promoted for the absence of glacial features. Unlike the surrounding territory, an area incorporating parts of Wisconsin, Minnesota, Iowa and Illinois is not covered by glacial sediments, and since debris deposited by glaciers is often referred to as glacial 'drift', this area is referred to as the 'Driftless Area'. It has a somewhat different topography and biogeography from surrounding areas by virtue of its non-glacial history. The official Iowa travel guide extols the virtues of the 160-km (100-mi.) Driftless Area Scenic Byway, which 'zigzags its way across the distinctive landscape of Allamakee County . . . As the last glaciers passed over Iowa, this corner of the state remained untouched, resulting in a striking region of gashed and furrowed terrain.'[5] If you are promoting tourism in Wisconsin or Iowa, having had no truck with ice-age glaciers can be presented as a very good thing. Their literature almost makes me feel sorry for poor old Switzerland.

Our relationship with glaciers has changed over time. The remote and hostile has to some extent been tamed and captured. What we once feared, we now have to protect from the threat that we ourselves present. So many of us have sought wilderness adventure and solitude that wilderness and solitude have become increasingly hard to find. Whereas in the past we encountered our adventures in the line of duty and exploration, now we construct adventures for their own sake. For many of us, glaciers and glacial environments provide an ideal setting for such adventures, however we may wish to take them. There are still opportunities to encounter deep wilderness and genuine danger, but for the overwhelming majority of us now glacier adventures are packaged and presented safely, such as from the window of the Glacier Express, or tamed when we are allowed to step down safely for a few moments from the Monster Truck Glacier Adventure onto a small section of glacier that has been inspected for safety.

For many adventure travellers, including those of us who choose to encounter glaciers, one of the attractions is the notion that we are entering into some kind of terra incognita, the idea that we are stepping where nobody has stepped before. In most of our world, even in the most remote deserts or on the highest mountains, there really is no unknown territory any more. However, glaciers are constantly changing. Day by day, year by year, they advance and retreat, covering or exposing new ground. Crevasses open and close, lakes form and drain. Camping on a glacier surface, the adventurer opens the tent flap each morning uncertain as to how the landscape might have changed overnight. Especially in this era of rapid environmental change and widespread glacier retreat every visit to a glacier is essentially a visit to a new landscape. One of the first glaciers that I knew well, Sólheimajökull in Iceland, has retreated many hundreds of metres in the few years since I last visited, and is unrecognizable in photographs that friends and colleagues show me of their visits. One of the tourism organizations that runs trips to Sólheimajökull has added a note to their website advising clients that because of global retreat they have combined two of their previous adventure and exploration walks into one, and that the glacier is now a forty-minute walk from the car park that used to be immediately adjacent to the ice. When next I visit, it will be like visiting a new glacier. No other landscape changes as much, or offers so many opportunities for constant rediscovery, as a glacier surface and glacier margin. In the short term this is exciting. In the long term, for all sorts of reasons that we will discuss in the final chapter, it is terrifying.

10 Glaciers and the Future

Glaciers are high on the environmental agenda because they tell us about the past, they are the canary in the coal mine for current dangers and they will be key components of future environmental change. The future of the whole planet, and all of us living on it, is tied up with the future of glaciers.

For a long time after scientists became convinced that human activity was the chief cause of present-day environmental change, even after the Intergovernmental Panel on Climate Change pronounced a scientific consensus, governments and businesses with vested interests in the chief causes of CO_2 pollution continued to deny what Al Gore calls the 'inconvenient truth' that human activities such as burning fossil fuel cause climate change that will lead to the melting of glaciers. In 2015 a climate-change workshop organized by the Vatican issued a statement that 'Human-induced climate change is a scientific reality, and its decisive mitigation is a moral and religious imperative for humanity.' Two months later Pope Francis issued an encyclical letter on 'Care for our common home', referring to the importance of glaciers for the Earth and the future of humanity, and warning of the dangers of melting polar ice caps.[1] The pope's message refers to water poverty, food security, rising sea levels and human migrations as serious issues associated with likely changes to glacier systems. The Church's engagement with the climate change debate was a significant development. By 2016 it was increasingly clear that only political and economic forces with vested interests in the status quo were still denying

Sadly Montana's Glacier National Park is now famous for the fact that its few remaining glaciers are rapidly disappearing.

the importance of climate change or the role of human activity in it. However, the United Nations Climate Change Conference at the end of that year confirmed that there remained substantial gaps in our commitment to combat climate change. In 2016 Dr Jérôme Chappellaz of the French National Centre for Scientific Research (CNRS) in Grenoble embarked on a project to start flying samples of ice from European glaciers to be placed in deep storage in Antarctica, so that the ice will be available for research after the glaciers themselves have disappeared.

The impact of climate change on glaciers, and the chain of events likely to follow from that, are among the most significant issues facing humanity. The most likely scenario for the immediate future is that mountain glaciers will shrink rapidly and that many will disappear within the lifespans of people who are alive now. This will have implications for water supply, sea-level rise and the sustainability of agriculture and habitation in many areas. Initially flooding will occur as glaciers melt. This will develop into drought and water shortages after the ice is gone. The longer-term future is less certain, but is likely to involve the disappearance of the major ice masses. In the most extreme possible future scenario, were the Antarctic Ice Sheet to melt completely, global sea levels would rise by as much as 50 or 60 m (160–200 ft). Around a billion people live within the areas that

An aerial photograph annotated to show the changing position of the perimeter of the retreating Sperry Glacier, in Montana's Glacier National Park: 1966, 1998, 2005, 2015.

would be inundated. This outcome would be well beyond the lifespans of the people who are making decisions about the environment today, but today's decisions will affect that long-term outcome as well as events in the short term. Even now, visible changes occurring to glaciers point directly towards a frightening future.

Glaciers will play a central role not only in the environmental drama of climate change and sea-level rise, but in accompanying political and economic dramas of environmental migration, resource pressure and international conflict. The long-term future calls up scenarios of new lands emerging from beneath disappearing ice sheets, old continents being overwhelmed by meltwater, the planet's major glacier-fed rivers drying up and huge swathes of densely populated agricultural land becoming desert. Millions of people become refugees – climate-change migrants. The potential upheaval is on an apocalyptic scale. In science fiction, future wars are based on possibilities such as the rise of the machines, or the invasion of aliens from another planet. A more likely future is one of water wars – essentially glacier wars – as nations fall into conflict over global shifts in the availability of water and land in the face of disappearing

A scientist from the U.S. Geological Survey captures images of the Grinnell Glacier in Glacier National Park as part of a repeat-photography project to illustrate glacial recession due to impacts of climate change.

glaciers, a drying of glacier-fed water supplies and the vanishing of land beneath the sea.

This has already begun.

The Intergovernmental Panel on Climate Change (IPCC) noted in 1990 that the greatest single impact of climate change could be on human migration. In 2008 the International Organization for Migration in their report 'Migration and Climate Change' predicted that by as soon as 2050 there would be 200 million climate-change migrants. Large-scale migration is recognized by the United Nations as a primary threat to international security. Even a relatively small sea-level change of a few metres, which could follow from the collapse of just part of the Antarctic Ice Sheet, would displace hundreds of millions of people. The majority of the world's largest cities are coastal, and much of the most densely populated agricultural land is on flood plains, deltas and estuaries close to sea level.

The Pacific island nation of Kiribati is already living in this future. In 2008, as it became clear that Kiribati was likely to be the first nation to become unsustainable in the face of rising water levels, the Kiribati government asked New Zealand and Australia to accept Kiribati citizens as permanent refugees. In 2012 the government purchased more than 2,000 hectares (4,900 ac) of land on the island of Fiji as part of a plan to cope when their own land is lost. By 2013 the government was urging its citizens to consider emigration and relocation. The population at that time was over 100,000. Some research has argued that coral growth and coastal sedimentation, combined with land reclamation, could save the islands. But even if the land itself remains above water level, problems such as the incursion of saltwater into underground aquifers will probably make the islands uninhabitable. If some of the land is lost and some saved, population density will rise to unsustainable levels on the areas that remain.

Apart from long-term sea-level rise, one of the most serious threats arising from the loss of glaciers is the concomitant loss of water supply. The disappearance of snow and ice from mountain summits has been given many different names around the world.

Many people refer to 'darkening peaks' as snow and ice vanish to reveal bare rock. In the Himalayas, this is often referred to as the 'black cloud'. The Tibetan Plateau and its surroundings contain the largest number of glaciers outside the polar regions. It is sometimes called 'the third pole', and the Hindu Kush Himalayan region is known as the Water Tower of Asia. The meltwater from Himalayan snow and ice feeds ten large river systems of South Asia: the Amu Darya, Brahmaputra, Ganges, Indus, Irrawaddy, Mekong, Salween, Tarim, Yangtze and Yellow Rivers. Snow and glacier melt is estimated to contribute more than 50 per cent of the total flow to the Indus river system. The Kathmandu-based International Centre for Integrated Mountain Development (ICIMOD) describes the glaciers of the Himalayas as 'nature's renewable storehouse of fresh water from which hundreds of millions of people downstream have benefited for centuries at the time in the year when it is most needed – the hot, dry season before the monsoon'.[2] ICIMOD estimates that these glacier-fed rivers are used as renewable sources of irrigation, drinking water and energy by 1.3 billion people. The Gangotri Glacier in the Himalayas, a major source of the water in the Ganges River, is retreating 35 m (115 ft) yearly. Once the glaciers disappear, rivers such as the Ganges will become seasonal, depriving hundreds of millions of people of perennial water.

Glaciers in the French Pyrenees are particularly vulnerable to climate change as they are so small, at such low elevations and at such low latitudes. The Ossoue Glacier, which is one of the largest glaciers in the region, has lost more than half of its area in the last hundred years and is predicted to disappear completely by the middle of this century. The future for Pyrenean glaciers looks bleak. In California, the glacier and snow cover of the Sierra Nevada has halved in the last 150 years. In the Colorado Rockies 42 per cent has been lost. In Montana's Glacier National Park glaciers and snow cover have dwindled by nearly 70 per cent. Turkey lost half of its glaciers between 1970 and 2013. The glaciers around Mount Everest are predicted to shrink by more than 70 per cent by the end of the current century. In East Africa, since 1900 the glaciers of Mount Kenya, Mount

The mouth of the Gangotri (Gaumukh) glacier: source of the River Ganges.

Kilimanjaro and the Ruwenzori Range have lost more than 80 per cent of their area. These glaciers are now almost extinct, and are predicted to be lost entirely within a few decades.

In Bolivia, the Chacaltaya Glacier was once the highest ski resort on Earth. In 1998 glaciologist Edson Ramirez predicted that the glacier would disappear completely before 2015. It had disappeared by the spring of 2009. Nobody goes summer skiing at Chacaltaya any more. Between 1975 and 2006 Bolivia's Cordillera Real lost nearly half the area of its glaciers. The reservoir in the Tuni Condoriri Range that provides 80 per cent of the drinking water to 2 million people living in the cities of La Paz and El Alto depends upon glaciers for more than half of its water supply, and these glaciers are forecast to disappear before 2050.

Hydrologist Pierre Chevallier and his colleagues published a detailed review in 2010 of how environmental change would affect glaciers and water resources in the tropical Andes, which are home to 99 per cent of the world's tropical glaciers.[3] Focusing on the Rio Santa Valley in Peru as a case study, they found that glacier water is of social and economic importance not only for the region but for the country as a whole. They distinguished different uses of glacier resources at different elevations. Above 5,000 m (16,400 ft) the glaciers and the mountain summits above them are a tourist asset which have attracted mountaineers from around the world for several decades. Between 2,000 and 4,000 m (6,500 and 13,000 ft), irrigated slope agriculture

Ossoue Glacier, one of the near-extinct glaciers of the Pyrenees.

has been practised for centuries with the help of complex channel systems. Below 2,000 m the waters of the Rio Santa drive turbines to generate electricity. Below 800 m (2,625 mi.), at the foot of the Andes, water from the Rio Santa is used to irrigate huge agricultural areas recently created in the barren coastal zone. As the glaciers disappear, all these resources will be lost. It has been estimated that the average yearly energy output of the Cañón del Pato hydropower plant on the Rio Santa would drop by 35 per cent once the glaciers' contribution to the water supply disappears, with an economic cost of U.S.$144 million per year. Since the 1990s, tourist visits to the Pastoruri Glacier have dwindled from 100,000 per year to less than one-third of that number as the glacier itself has diminished.

The progressive loss of mountain glaciers is creating a range of environmental hazards, such as the creation of unstable ice-marginal lakes that can burst, creating devastating floods. In the middle of the twentieth century, the Cordillera Blanca witnessed a series of major floods caused by the formation and catastrophic drainage of ice-marginal lakes. In the biggest flood, in 1941, Lake Pallqaqucha burst its dam, causing a flood and mudflow that killed around 6,000 people. Following that disaster and several similar events, the Peruvians embarked on a major programme to develop engineering solutions to this threat. Glacier-monitoring systems were put in place and lakes were artificially drained or limited. The experience of the Cordillera Blanca is now being repeated in the Himalayas and elsewhere, as glacier retreat becomes a problem for mountainous regions around the world. For example, in the Dudh Kosi, the largest glacierized basin in Nepal, nearly all of the valley glaciers are retreating, some at rates of up to 70 m (230 ft) each year. This has led to the formation of dozens of new lakes. In remote, inaccessible areas at such high altitude it is a major challenge even to monitor these lakes as they grow, let alone put in place engineering solutions to the hazard. Even satellite remote sensing is not always effective, as water can build up unseen underneath or inside a glacier. Glaciologist Doug Benn has described some Himalayan glaciers as being like Swiss cheese, full of holes. One of these dangerous glacier lakes developed in front of the Ngozumpa Glacier in the Dudh Kosi where Benn and his colleague Sarah Thompson set up a network of remote cameras to monitor changes in the glacier.[4] Initially the lake grew slowly, but from 2001 it expanded by about 10 per cent every year. As the glacier continues to retreat, it is predicted that the lake could attain a volume of hundreds of millions of cubic metres within a few decades. If its dam collapses, the resulting flood will be catastrophic.

Efforts to address the threats posed by glacier retreat have been made at a number of different scales. At the largest scale are attempts to limit or reverse the human impact on global climate. At the smallest scale, many projects have been established to counteract or reverse the effects of warming on specific glaciers. Some projects aim to artificially protect the water supply provided by glaciers. Melting can be increased or reduced by spreading different materials on top of snow and ice. Traditional farming methods in parts of Asia have involved spreading dirt on spring snow to speed its melting, and a similar procedure can be used to increase summer melting of glaciers. A thin layer of a dark, heat-absorbing material such as powdered coal can increase surface melting by as much as 55 per cent, while a thick coating, or use of a thermally insulating material, can reduce melting. Another approach has been to grow what are sometimes referred to as 'artificial glaciers'. Different methods of creating ice artificially have been described. One, sometimes called glacier grafting, involves moving blocks of ice from an already existing glacier to a shaded high-altitude hillside where they may survive without melting. The ice blocks are covered with earth or rocks, and groundwater or rainfall is allowed to freeze around them. Although these are not glaciers in the technically correct sense of the word, they can provide a similar year-round water supply in arid, mountainous regions. 'Glaciers' grown in this way were used in the thirteenth century as a means of blocking and defending mountain valleys in what is now northern Pakistan. The method has been practised for centuries as a way of improving water supply to villages where snowmelt ran out before the end of the growing season. Now, as remote mountain communities come under pressure from climate change, this approach is being considered as a partial response to the problem of glacier retreat. In the cold trans-Himalayan desert of Ladakh, where settlement depends on the use of glacier meltwater for irrigation, water shortages are particularly acute in April and May, at the start of the growing season but before the main meltwater season, when water supply is low but demand is high. A project established by The Students' Educational and Cultural Movement of Ladakh (SECMOL) involves freezing surplus water that flows in the winter when demand is low, and storing it as ice until it is needed in the spring. Driven by gravity, water is diverted through pipes and released several metres above ground at a location where, falling through the cold air, it will freeze as it reaches the ground below, forming a cone, or stupa, of ice. Prototypes have demonstrated that these 'ice stupa' could potentially store substantial amounts of winter water for release at the critical drought period in the spring.

A similar problem in a different economic setting has seen efforts to preserve or rebuild existing glaciers by means of snowmaking machines. Theodulgletscher, in Switzerland's Zermatt ski resort, is one of the most visited glaciers in the world. Each winter some 2 million people go skiing in Zermatt. In response to glacier retreat, snowmaking machines are being used to add extra snow onto their surfaces and extend the ski runs. Zugspitze, Germany's highest mountain, has parts of its snowfield covered each summer with reflective tarpaulins to protect winter snow from melting so it can contribute to the glacier. It is estimated that about 80,000 cubic metres of snow is saved each year. At Italy's Presena Glacier, 90,115 square metres of 4-mm-thick insulating material was spread over the ice in an effort to protect it. In the short term, at least, the method was successful: melting underneath the insulation was up to 60 per cent less than in uncovered areas. At Whistler Blackcomb in British Columbia, North America's largest ski resort, artificial snowmaking was put into action in 2015. The park's mountain planning and environmental resource manager told the press, 'This is a first in the industry where a mountain resort is looking at building a large enough snowmaking system that can reverse a retreat of a glacier.'[5] Glaciologists asked to comment on the plan were doubtful about its long-term success. One limitation to both snow canons and insulating covers is their cost, and while it may be economically viable for a few small resort areas, the system has yet to be adopted on any large scale that would be significant for the global crisis of glacier extinction.

Not all of the problems associated with glacier retreat are so immediately obvious. A side-effect of glacier retreat in Huascarán Biosphere Reserve in Peru's Cordillera Blanca has been the exposure to the atmosphere of metal-rich rocks that have been covered by ice for tens of thousands of years. The newly exposed rocks are susceptible to rapid weathering, and the chemical breakdown of the rocks has caused discharge of acidic water into the local ecosystem, carrying lead, arsenic and cadmium into the water supply. In response, carefully selected plant species that have the capacity to absorb the pollution are being planted.

Despite the overwhelmingly negative implications of disappearing glaciers, there are also positives to be found. One potentially massive economic benefit of glacier retreat is the progressive exposure of land suitable for mineral exploitation. In areas such as Greenland, where enormous mineral reserves are likely to exist in areas that are presently beneath the ice, glacier retreat equates to an opening frontier of mineral exploration. Of course, the enthusiasm of geologists who relish the opening up of new areas for exploration has to be weighed against the enthusiasm of those a few decades ago who recommended that it would be safe to bury our radioactive waste underneath glaciers. Had that plan gone forward, our view of the future might have included the radioactive fragments of Earth's last glaciers melting into a radioactive ocean 150 m (490 ft) deeper than today.

Glaciers: a future beyond physical geography

The way we look at glaciers has changed over time. Part of this change has involved a recognition that glaciers play an important role in the interconnected global system, and that the glaciers, and the system, are themselves changing. The development of our environmental understanding of glaciers has also led to a change in their cultural status. Glaciers have become symbols both of pristine wilderness and of human-induced environmental change. This is reflected in the way glaciers are treated in art as well as in science. We must allow our knowledge of glaciers and the global system to affect the way we treat the environment and think about the economy. Just as science and art are linked, science must be linked with the economics and the politics of addressing environmental threats. What can be done to prevent glaciers melting in a warmer climate? What can be done to prevent sea-level rise? What can be done to prevent the catastrophic floods associated with some stages of glacier decay? What can be done to prevent the intercontinental migrations and international conflict – the glacier wars – which could follow the combined drought and coastal flooding associated with major glacier retreat? These are not only questions for science, but human, social questions. A plan for a sustainable future that will ward off the threat of glacier conflict will not be the plan of scientists, or artists, or politicians alone. We need visions that embrace all of these points of view. The ice age in which we live is not just an age where glaciers affect the landscape, but one where they affect our whole way of looking at the world.

We need not just one prospectus for the future of glaciers, but several. For glacier science we need a prospectus that includes the changes in glaciers associated with future climate change. Glacier science must also embrace glaciers on other planets, of which we so far understand very little and from which we might have much to learn. We also need a prospectus for glaciers and the arts, in which one role of art is to inspire science, and another is to communicate with people, and in ways that science cannot. Finally we need a prospectus for glaciers and society that engages politics and economics with the hazards, resources and potential conflicts associated with glaciers in a changing environment.

Physical geography, the study of the Earth's surface and near-surface environments, connects us to something beyond ourselves. It is not only about the physical environment, but about our human relationship with it. We need more-than-physical geographies that treat the simply physical as a link to something more. Our geographies must recognize that having glaciers in the world not only makes our world different, but makes us different. That difference stems not only from what glaciers are, but from how we imagine and represent them. The geographical imagination is not just an archive of past thought, memory and ideas but a structuring of our vision of how we will be in the future.

Science has been through a period of several hundred years of disciplinary segregation. This was a substantial change from earlier pre-disciplinary eras, in which human enquiry was much less compartmentalized. Geographers, historians, chemists and engineers today inhabit different departments and publish in different journals. Interdisciplinary collaborations are heralded as unusual innovations in practice. The same is true for academics, artists and politicians. We have stayed too long behind our territorial fences. But there are signs that we are starting to move towards a post-disciplinary future, and it is in the area of environmental risk that we are most likely to make those connections. That is one of the most obvious intersections where geographies encounter economics, politics, engineering, planning and environmental management.

Some things are unimaginable – eternity, the end of the world. Most of us don't really engage with those things. Other things are trivial, and make up most of our daily lives. Glaciers lie somewhere between the two. On the one hand they are graspable, real-world objects that we can visit and touch, and even if we don't visit them we can easily form an image of them in our mind. But on the other hand they are connected to the vast, the sublime, the other. They give us a link, a glimpse, a way of approaching the unimaginable: a way of seeing more, beyond the obvious, through the boundary between vision and imagination. Glaciers are a reminder and relic of something we could have forgotten. Humanity has seen only a tiny fraction of history and so has witnessed only some of the things that have happened, that can happen, and that will happen again in the physical environment. We missed the Big Bang, obviously, and we don't remember the origins of life. We missed the dinosaurs and we haven't seen any big meteorite impacts. We missed the Altai and Missoula floods. We nearly missed the glaciers. The Earth's remaining glaciers are like mammoths in the zoo, linking us to a past that has largely disappeared and that we could have entirely forgotten. Of course we have no mammoths in the zoo. But we do still have glaciers, for a while. As a scientist I use the mammoth analogy, and regret that I couldn't have witnessed the Missoula floods. Were I not a scientist I might cite not the mammoth but the unicorn. There are no unicorns to link us to that metaphysical imaginary magical world. But, for those of us

At Skaftafellsjökull in Iceland, the decaying glacier leaves behind a complex landscape that includes a series of moraine ridges indicating successive positions of the retreating margin.

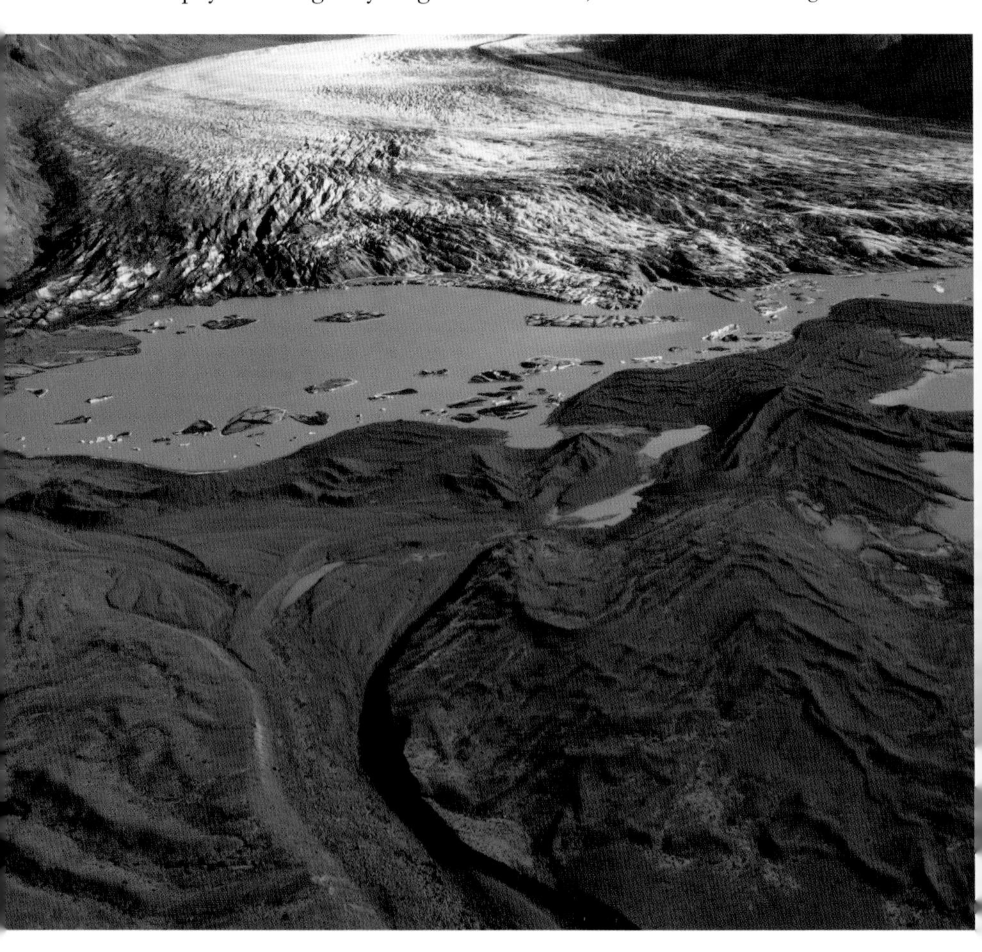

High above the glaciated mountain landscape of the European Alps, the Cygnus commercial cargo craft approaches the International Space Station. The view from Space gives scientists a new perspective on the Earth's surface, and glaciers continue to be a major part of our future.

who are so inclined, glaciers can do the same. They are grander than the normal scope of human comprehension. They enable us to see the world in a way, in a context, at a scale that we would otherwise struggle to notice.

SOME NOTABLE GLACIERS

A selection of interesting glaciers which are worth looking up and learning more about. Most of these can be searched by name on Google Earth or Google Maps.

Aletsch Glacier, Switzerland
The largest glacier in the European Alps

Athabasca Glacier, Canada
An easily accessible glacier, popular with tourists, in British Columbia

Baltoro Glacier, Pakistan
One of the longest glaciers outside the polar regions

Bering Glacier, Alaska, USA
The largest glacier in continental North America

Breiðamerkurjökull, Iceland
Famous for its glacier lagoon, and frequently featured in film and TV

Castle Creek Glacier, Canada
A British Columbia glacier notable for its clear sequence of recessional moraines

Chacaltaya Glacier, Bolivia
Iconic symbol of the dangers of climate change; this glacier disappeared in 2009

Columbia Glacier, Alaska, USA
Famous for rapid retreat of its floating terminus in the late 20th and early 21st centuries

Crater Glacier, Mt St Helens, USA
The glacier regrew inside the volcano's crater after the eruption in 1980

Drangajökull, Iceland
Iceland's northernmost glacier

Drygalski Ice Tongue, Antarctica
A floating extension of the David Glacier, which protrudes into the Ross Sea

Elephant Foot Glacier, Greenland
Spectacular example of a piedmont lobe: a glacier margin spreading out into a rounded shape

Erebus Ice Tongue, Antarctica
The floating terminus of the Erebus Glacier, which extends into McMurdo Sound

Fedchenko Glacier, Tajikistan
One of the longest valley glaciers outside the polar regions

Fox Glacier and Franz Josef Glacier, New Zealand
Two easily accessible glaciers, popular with tourists, within the Te Wāhipounamu World Heritage Site

Gangotri Glacier, India
One of the largest glaciers in the Himalayas, and a primary source of the Ganges River.

Hardangerjøkulen, Norway
Scenes for the film *Star Wars: The Empire Strikes Back* were shot here

Hubbard Glacier, Alaska, USA
Has periodically blocked the outlet of Russell Fjord to create a temporary lake

Jakobshavn Isbrae, Greenland
Often cited as the fastest-flowing glacier in the world. May have spawned the iceberg that sank the *Titanic*

Jostedalsbreen, Norway
The largest continuous ice mass in continental Europe

Khumbu Glacier, Nepal
Part of the trail to one of the Everest base camps, this is the world's highest glacier

Kongsvegen and Kronebreen Glaciers, Svalbard, Norway
Two well-studied surging glaciers, which merge into a single floating terminus

Lambert Glacier, Antarctica
An ice stream within the Antarctic Ice Sheet, often cited as the world's longest glacier

Lewis Glacier, Kenya
One of the most closely monitored tropical glaciers, it is disappearing rapidly

Lyell Glacier, Yosemite, USA
The largest glacier in Yosemite when it was discovered, the Lyell has shrunk dramatically and stopped moving, so it is no longer technically a glacier

Lysii Glacier and Davidov Glacier, Kyrgyzstan
Mining activity including excavation of ice and dumping of rock onto the ice has substantially altered the behaviour of these two glaciers

Malaspina Glacier, Alaska, USA
The biggest mountain glacier outside Antarctica, with a spectacular piedmont lobe

Matanuska Glacier Alaska, USA
A well-studied glacier, accessible by road from the Glenn Highway

Mer de Glace, France
Near Mont Blanc, this valley glacier is famous for its association both with the Romantics and with early glacier science

Muir Glacier, Alaska, USA
This glacier in Glacier Bay National Park has suffered rapid, well-documented retreat

Mýrdalsjökull, Iceland
An ice cap from beneath which volcanic eruptions have caused major floods

Nigardsbreen, Norway
A valley or outlet glacier descending from the Jostedalsbreen Ice Cap

Perito Moreno Glacier, Argentina
Due to its size, accessibility and calving icebergs this glacier is a major tourist attraction in the Los Glaciares National Park

Pine Island Glacier, Antarctica
This glacier is thought to contribute more ice to the sea than any other

Pio XI glacier (also known as Brüggen Glacier), Chile
A major outlet of the Southern Patagonian Ice Field, this glacier is unusual in that it experienced some advance in the late 20th century

Quelccaya Ice Cap, Peru
The largest glaciated area in the tropics, famous for scientific work on long-term climate change revealed in its ice

Renland Ice Cap, Greenland
An ice cap situated on a high-elevation plateau

San Rafael Glacier, Chile
A major outlet glacier of the Northern Patagonian Icefield, and closer to the equator than any other tidewater (sea-level) glacier

Siachen Glacier, Karakoram Himalaya
The focus of a territorial dispute and military action between India and Pakistan

Sólheimajökull, Iceland
Popular on the tourist trail, but recent retreat has made it less accessible

Svínafellsjökull, Iceland
One of several easily accessible glaciers in the Vatnajökull National Park

Tasman Glacier, New Zealand
The longest glacier in New Zealand, currently in rapid retreat and predicted to disappear within a few decades

Taylor Glacier, Antarctica
A long, cold-based glacier draining from the Antarctic Ice Sheet

Thwaites Glacier, Antarctica
A broad, fast-flowing glacier, thought to be particularly vulnerable to retreat in response to ongoing environmental change

Variegated Glacier, Alaska, USA
Famous for an extended period of scientific research on glacier surges

Vatnajökull, Iceland
A large ice cap, beneath which are a number of active volcanoes

Vernagtferner, Austria
A particularly well-documented glacier, with records of glacier-related floods since the 16th century

Zachariae Isstrom, Greenland
Recent rapid acceleration of this ice stream is thought to pose a threat to the stability of the Greenland Ice Sheet

REFERENCES

1 Ways of Thinking about Glaciers

1 J. G. Cogley et al., *Glossary of Glacier Mass Balance and Related Terms* (Paris, 2011).
2 Ted Kooser, 'After Years', in *Solo: A Journal of Poetry* (Carpinteria, 1996).
3 Mark Carey, 'The History of Ice: How Glaciers Became an Endangered Species', *Environmental History*, 12 (July 2007), pp. 497–527.
4 Julie Cruikshank, *Do Glaciers Listen? Local Knowledge, Colonial Encounters, and Social Imagination* (Vancouver, 2005), p. 3.
5 Brian M. Fagan, *The Little Ice Age: How Climate Made History, 1300–1850* (New York, 2000).
6 Robert E. Rhoades, Xavier Zapata Ríos and Jenny Aragundy Ochoa, 'Mama Cotacachi: History, Local Perceptions, and Social Impacts of Climate Change and Glacier Retreat in the Ecuadorian Andes', in *Darkening Peaks: Glacier Retreat, Science and Society*, ed. Ben Orlove, Ellen Wiegandt and Brian H. Luckman (Berkeley, CA, 2008), pp. 216–25.
7 H. J. Zumbühl, D. Steiner and S. U. Nussbaumer, '19th Century Glacier Representations and Fluctuations in the Central and Western European Alps: An Interdisciplinary Approach', *Global and Planetary Change*, 60 (2008), p. 44.
8 Carey, 'The History of Ice'.
9 University of Leeds, 'A Discipline of the Mind: The Drawings of Wilelmina Barns-Graham', *Arts and Culture News*, 3 November 2009, www.leeds.ac.uk.
10 W. G. Hoskins, *The Making of the English Landscape* (London, 1955).
11 Siân Ede, *Art and Science* (London, 2005).
12 David Sugden and Brian John, *Glaciers and Landscape* (London, 1976).

13 Richard B. Alley, 'Flow-law Hypotheses for Ice-sheet Modelling', *Journal of Glaciology*, 38 (1992), p. 245.
14 Richard B. Alley, *The Two-mile Time Machine* (Princeton, NJ, 2000).

2 How Glaciers Work

1 Stewart S. R. Jamieson et al., 'An Extensive Subglacial Lake and Canyon System in Princess Elizabeth Land, East Antarctica', *Geology*, XLIV/2 (2016), pp. 87–90.
2 P. Rastner et al., 'The First Complete Glacier Inventory for the Whole of Greenland', *Cryosphere Discussions*, VI/4 (2012), pp. 2399–436.
3 D. R. MacAyeal, 'Binge/Purge Oscillations of the Laurentide Ice Sheet as a Cause of the North Atlantic's Heinrich Events', *Paleoceanography*, VIII/6 (1993), pp. 775–84.
4 Richard P. Feynman, *The Character of Physical Law* (London, 1965).

4 A Short History of Glacier Science

1 Louis Agassiz, *Études sur les glaciers* (Neuchâtel, 1840).
2 F. Darwin, ed., *The Life and Letters of Charles Darwin* (London, 1887), vol. I, p. 25.
3 J. G. Cogley et al., *Glossary of Glacier Mass Balance and Related Terms* (Paris, 2011).
4 *Glacialist's Magazine*, I/I (August 1893), p. 22.
5 Edward B. Evenson et al., 'Enigmatic Boulder Trains, Supraglacial Rock Avalanches, and the Origin of "Darwin's Boulders", Tierra del Fuego', *GSA Today*, XIX (2009), pp. 4–10.
6 John Ruskin cited in Edward Tyas Cook, *The Life of John Ruskin*, vol. II: *1860–1900* (1911), p. 59.
7 T. G. Bonney, *The Story of our Planet* (London, 1893), p. 154.
8 Grove Karl Gilbert, *Glaciers and Glaciation*, Alaska Series, vol. III (New York, 1904).
9 Louis Lliboutry, 'Monolithologic Erosion of Hard Beds by Temperate Glaciers', *Journal of Glaciology*, XXXVI (1994), pp. 433–50.
10 W. L. Rogers, 'The Philosophy of Glacier Motion', *Journal of the American Geographical Society of New York*, XX (1888), pp. 481–501.
11 Richard B. Alley, 'Flow-law Hypotheses for Ice-sheet Modelling', *Journal of Glaciology*, 38 (1992), pp. 245–56.
12 Neil F. Glasser et al., 'Terrestrial Glacial Sedimentation on the Eastern Margin of the Irish Sea Basin: Thurstaston, Wirral', *Proceedings of the Geologists' Association*, 112 (2001), pp. 131–46.

13 Arthur R. Dwerryhouse, 'Notes on the Glacial Deposits on the Cheshire Shore of the Dee Estuary', *Glacialists' Magazine*, 2 (1895), pp. 150–57.
14 David Sugden, 'Changing Glaciers and Their Role in Earth Surface Evolution', in *Glacier Science and Environmental Change*, ed. Peter G. Knight (Oxford, 2006), pp. 188–91.
15 P. Rastner et al., 'The First Complete Glacier Inventory for the Whole of Greenland', *Cryosphere Discussions*, VI/4 (2012), pp. 2399–436.
16 W. Tad Pfeffer et al., 'The Randolph Glacier Inventory: A Globally Complete Inventory of Glaciers', *Journal of Glaciology*, 60 (2014), pp. 537–52.

5 Glaciers and the Big Global System

1 B. Marzeion and A. Nesje, 'Spatial Patterns of North Atlantic Oscillation Influence on Mass Balance Variability of European Glaciers', *The Cryosphere*, 6 (2012), pp. 661–73.
2 Richard B. Alley, *The Two-mile Time Machine* (Princeton, NJ, 2000).

6 Glacier Economics: Hazards, Resources, Politics

1 Jennifer Cox, 'Finding a Place for Glaciers within Environmental Law: An Analysis of Ambiguous Legislation and Impractical Common Law', *Appeal*, 21 (2016), pp. 2–36.
2 Mark Carey, *In the Shadow of Melting Glaciers: Climate Change and Andean Society* (New York, 2010), p. 7.

7 Glaciers in Art

1 Letter from W. Barns-Graham to Tate Gallery, 2 February 1965, Tate Gallery cataloguing file for *Glacier Crystal, Grindelwald*, cat. no. T00708, www.tate.org.uk/art/artworks.
2 Gil Docking and Michael Dunn, *Two Hundred Years of New Zealand Painting* (Auckland, 1990), p. 54.
3 CBC TV, *The Group of Seven: The Myth of the Unspoiled Wilderness* (TV programme), first broadcast 22 February 1996. Available at www.cbc.ca/archives.
4 Grove Karl Gilbert, *Glaciers and Glaciation*, Alaska Series, vol. III (New York, 1904), p. 4.
5 Alexis Drahos, 'Brett's Boulders', *Geoscientist*, 19 (2009).
6 British Antarctic Survey (online), 'Data as Art', www.bas.ac.uk/project/data-as-art.
7 Elizabeth Jackson (online), 'Glacier Project', https://glacierproject.wordpress.com.

8 Samuel U. Nussbaumer and Heinz J. Zumbühl, 'The Little Ice Age History of the Glacier des Bossons (Mont Blanc Massif, France): A New High-resolution Glacier Length Curve Based on Historical Documents', *Climatic Change*, 111 (2012), pp. 301–34.
9 Siân Ede, *Art and Science* (London, 2005).
10 Anna McKee (online), *Encounters with an Ice Core*, http://annamckee.com, 18 December 2008.

8 Glacier Stories and Songs . . . Once upon a Glacier

1 M. Scott Peck, *The Road Less Travelled* (New York, 1978).
2 Snorri Sturluson, *The Prose Edda: Tales from Norse Mythology* (New York, 1916).
3 Hans Haid, *Mythos Gletscher* (Innsbruck, 2004).
4 Patrick D. Nunn and Nicholas J. Reid, 'Aboriginal Memories of Inundation of the Australian Coast Dating from More than 7,000 Years Ago', *Australian Geographer*, XLVII/1 (2016), pp. 11–47.
5 Rebecca Solnit, *A Field Guide to Getting Lost* (New York, 2005).
6 Julie Cruikshank, *Do Glaciers Listen? Local Knowledge, Colonial Encounters, and Social Imagination* (Vancouver, 2005).
7 Halldór Laxness, *Under the Glacier* (New York, 2007).
8 Mary Shelley, *Frankenstein, or, The Modern Prometheus* (London, 1818).
9 Charles Isaac Elton, *An Account of Shelley's Visits to France, Switzerland, and Savoy, in the Years 1814 and 1816* (London and Edinburgh, 1894).
10 Lord Byron, *Childe Harold's Pilgrimage* (London, 1816).
11 Odysseus Elytis, *The Axion Esti*, trans. Edmund Keeley and George Savidis (London, 1980).
12 Charles Augustus Keeler, 'To an Alaskan Glacier', in *Idylls of El Dorado* (San Francisco, CA, 1900).
13 Email communication with author, 29 August 2011.
14 Email communication with author, 16 February 2019.
15 W. G. Hoskins, *The Making of the English Landscape* (London, 1955).

9 Adventure, Exploration, Inspiration

1 Halldór Laxness, *Heimsljós* (World Light) [1940], trans. Magnus Magnusson (London, 2002).
2 See www.icehorizons.com, accessed 10 February 2019.
3 Jim Ring, *How the English Made the Alps* (London, 2000).
4 See www.zermatt.ch/en, accessed 10 February 2019.

10 Glaciers and the Future

1 'Encyclical Letter *Laudato Si'* of the Holy Father Francis on Care of Our Common Home', available at http://w2.vatican.va, accessed 15 February 2019.
2 'Cryosphere', www.icimod.org, accessed 15 February 2019.
3 Pierre Chevallier et al., 'Climate Change Threats to Environment in the Tropical Andes: Glaciers and Water Resources', *Regional Environmental Change*, 11 (2010), pp. 179–87.
4 Sarah Thompson et al., 'A Rapidly Growing Moraine-dammed Glacial Lake on Ngozumpa Glacier, Nepal', *Geomorphology*, 145 (2011), pp. 1–11.
5 Vince Shuley, 'Horstman Glacier gets a Helping Hand', www.whistlerquestion.com, accessed 15 February 2019.

ASSOCIATIONS AND WEBSITES

Alpine Club
www.alpine-club.org.uk

Glacier Hub
http://glacierhub.org

Glaciers Online
www.swisseduc.ch/glaciers

GlacierWorks
http://glacierworks.org

GLIMS Global Land Ice Measurements from Space initiative
www.glims.org

GTN-G Global Terrestrial Network for Glaciers
www.gtn-g.ch

IACS International Association of Cryospheric Sciences
www.cryosphericsciences.org

IGS International Glaciological Society
www.igsoc.org

NSIDC U.S. National Snow and Ice Data Center
http://nsidc.org

WGMS World Glacier Monitoring Service
http://wgms.ch

ACKNOWLEDGEMENTS

The author is indebted to many individuals for support, advice, information and resources received during the writing of this book, including, in particular, Peter Adey, Richard Allaway, Daniel Allen, Mia Baila, Roger Braithwaite, Peter Broderick, Miriam Burke, Mark Carey, Adriana Craciun, Gareth Digges La Touche, Sydney Fleck, Chris Fogwill, Zena and George Furby, Robin Garton, Jocelyn Hirose, Teresa Holligan, Debbie Knight, Ted Kooser, Magnús Már Magnússon, Keith Montgomery, Simon Ommanney, Mauro Rubino, Ned Selfe, Katie Szkornik, Hugh Torrens, Bert Ulrich (NASA), Jaap van der Meer, Richard Waller, Alice Witherick, Peter Worsley, Kathryn Yusuff and Heinz J. Zumbühl.

Special thanks, of course, to Debbie, not only for her research on this project but for being the best field assistant I ever worked with . . . and for still being my wonderful wife.

PHOTO ACKNOWLEDGEMENTS

The author and publishers wish to express their thanks to the below sources of illustrative material and/or permission to reproduce it. Some locations of artworks are also given below, in the interests of brevity:

© Mia Baila: pp. 82–3, 142; the British Museum: p. 134; © Miriam Burke: p. 151; Cornell University Library, Division of Rare and Manuscript Collections: pp. 77, 121, 122; Esercito Italiano, Ministero della Difesa: p. 114; © David Etheridge, courtesy Mauro Rubino: p. 63; Samuel Ferrara on Unsplash: p. 30; Fylkesarkivet i Sogn og Fjordane, Leikanger: p. 173; Institute for Svalbard and Ice Sea Exploration, Norway: p. 84; photos Peter G. Knight: pp. 35, 37, 41, 45, 47, 87, 88, 93, 117, 166, 185; Library of Congress, Washington, DC: p. 18; Maridav/Shutterstock.com: p. 6; The Metropolitan Museum of Art, New York: pp. 146, 148, 149, 156; photos NASA: pp. 8, 38, 105, 106, 199; photos NASA: pp. 22–3, 39, 42–3, 62, 96, 100–101, 102 (Earth Observatory), 91 (Chris Larsen), 40 (Michael Studinger), 92 (USAF 30th Space Wing/Timothy Trenkle); Nasjonalbiblioteket, Oslo: p. 97; Nasjonalmuseet for kunst, arkitektur og design, Oslo: p. 136; National Oceanic and Atmospheric Administration: p. 26; Claire Nolan on Unsplash: p. 28; Official White House Photo by Pete Souza: p. 113; photos Pixabay: pp. 10 (girlart39), 11 (jackmac34), 55 (OrcaTec), 12–13 (Simon), 56–7, 158–9, 165, 176 (Simon Steinberger), 48, 52–3, 178 (Skeeze); private collection: 137, 140–41; © Mauro Rubino: pp. 85, 110; courtesy Ted Scambos and Rob Bauer, NSIDC: p. 20; Schweizerische Nationalbibliothek, Bern, the Gugelmann Collection: p. 133; Bernard Spragg: p. 29; © Katie Szkornik: p. 90; © Gareth Digges La Touche: p. 125; © Prof Chris Turney, courtesy Chris Fogwill: p. 86; photos U.S. Air Force: pp. 34 (tech. Sgt. John Gordinier), 116 (Sgt. Anna-Marie Wyant); photos U.S. Geological Survey: pp. 67, 186, 187 (Lisa McKeon), 76 (Austin Post); © Richard Waller: pp. 36, 44, 46, 80, 103, 198; © Alice Witherick: p. 69; Yale University Art Gallery: p. 145; Zentralbibliothek, Zürich: p. 16.

INDEX

Page numbers in *italics* refer to illustrations